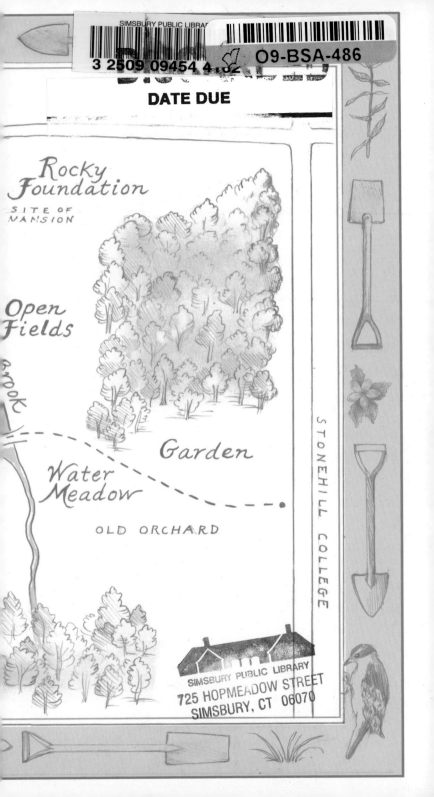

Rocky
Foundation

SITE OF
MANSION

Open
Fields

rook

Garden

Water
Meadow

OLD ORCHARD

STONEHILL COLLEGE

THE PATH

THE

PATH

A One-Mile Walk

Through the Universe

CHET RAYMO

WALKER & COMPANY

New York

First published in the United States of America in 2003 by
Walker Publishing Company, Inc.

Published simultaneously in Canada by Fitzhenry and Whiteside,
Markham, Ontario L3R 4T8

For information about permission to reproduce selections from
this book, write to Permissions, Walker & Company, 435 Hudson
Street, New York, New York 10014

Library of Congress Cataloging-in-Publication Data
available upon request
ISBN 0-8027-1402-1

BOOK DESIGN BY KATY RIEGEL
ENDPAPER MAP AND ILLUSTRATIONS BY MARTIE HOLMER

Visit Walker & Company's Web site at www.walkerbooks.com

Printed in the United States of America

2 4 6 8 10 9 7 5 3 1

CONTENTS

CONTENTS

ACKNOWLEDGMENTS

I thank family members, friends, and students who have shared the path with me over the years. Hazel L. Varella, a longtime acquaintance and Easton historian, was of inestimable help, especially as author, with Elise Ames Parker, of *Growing Up at Sheep Pasture* (Easton Historical Society, 1976). Hazel read an early draft of the manuscript and made important suggestions and corrections. Edmund Hands, another Easton historian, supplied me with valuable information, including census data for Jenny Lind Street. Michael Dosch at the Frederick Law Olmsted Archives in Brookline, Massachusetts, made it possible for me to examine relevant drawings of Olmsted's firm. Greg Galer, archivist of the Tofias Industrial Archives at Stonehill College, was generous with his time and knowledge. Paul Berry, curator of the Easton Historical Society, made available historical photographs.

ACKNOWLEDGMENTS

Books that were especially valuable to me include: Edmund Hands, *Easton's Neighborhoods* (Easton Historical Society, 1995); William L. Chaffin (1837–1923), *History of Easton, Massachusetts* (1886); Margaret McEntee, Edmund Hands, Jeffrey Nystrom, Duncan Oliver, Hazel Varella, and Robert Brown, *History of Easton,* volume 2 (Easton Historical Society, 1975); Peter J. Schmitt, *Back to Nature: The Arcadian Myth in Urban America* (Oxford, 1969); Norman T. Newton, *Design on the Land: The Development of Landscape Architecture* (Belknap Press of Harvard University Press, 1971); and Rebecca Solnit, *Wanderlust: A History of Walking* (Penguin, 2000).

Jackie Johnson, my editor at Walker & Company, walked the path and polished the book; I am privileged to enjoy her talent and friendship. George Gibson, my publisher at Walker, continues to believe in me. Thanks go to all the fine people at Walker and to my agent, John Williams. The Lannan Foundation provided support. My wife, Maureen, gave the manuscript her usual careful reading. And finally, thanks to the directors and staff of the Natural Resources Trust of Easton for preserving and maintaining the path and its environs for the enjoyment of walkers.

THE PATH

PROLOGUE

FOR THIRTY-SEVEN YEARS I have walked the same path back and forth each day from my home in the village of North Easton, Massachusetts, to my place of work, Stonehill College. The path takes me along a street of century-old houses, through woods and fields, across a stream, along a water meadow, and through an old orchard and community gardens. Much of the landscape of the path was designed by the famous American landscape architect Frederick Law Olmsted as an estate called Sheep Pasture for a great-grandson of Oliver Ames, the man who established in 1803 the shovel manufactory whose history is so intimately linked to the village. I have walked the path so many times, I believe I could do it blindfolded; certainly, I have done it on the darkest nights.

If it is possible to know a landscape well, I know this one. I can anticipate the exact day in late February when I will hear the first red-winged blackbirds taking up territories along the brook. I know to the hour when the spring peepers in the water meadow will begin their song, and when the wood anemones will open their five-petaled blossoms beside the shaded path. I know day by day the moments of sunrise and sunset, when the new moon will grace the western sky with its eyelash crescent, and when Orion will rise in the east. After thirty-seven years this knowledge is in my bones, put there by long experience, by close observation, by love. At the same time I know that it is not possible to know any landscape exhaustively. For all of its familiarity, there has never been a day I have walked the path without seeing something noteworthy. There are some things I have seen only once in all those years: a single blossom of wild columbine, a kingfisher by the stream, a dog stinkhorn mushroom.

Every pebble and wildflower has a story to tell. The flake of granite in the path was once at the core of towering mountains pushed up across New England when continents collided. The purple loosestrife beside the stream emigrated from Europe in the 1800s as a garden ornamental, then went wantonly native in a land of wild frontiers. The light from the star Arcturus I see reflected in the brook beneath the bridge at night has been traveling across space for forty years before entering my eye. I have attended to all of these stories and tried to hear what the landscape has to say. Binoculars and magnifier helped: binoculars for the red-tailed hawk at the top of the distant pine, magnifier to inspect the clever sexual parts—male paint brush and female sticky pad—of the cardinal flower.

Of course, no one person has the time, knowledge, or skill to learn everything about a landscape, so in my walks I have relied upon the labors of generations of botanists, ornithologists, zoologists, geologists, ecologists, meteorologists, astronomers, cultural historians, and a host of other specialists who have studied with particular care some feature of the natural world. Whenever possible, I queried people I met along the way: the old people who grew up in the landscape, who knew it in its former incarnations, who watched it change; and the children, who still have the capacity to see everything afresh and to see things the rest of us might miss. I have attended, too, to language. How did the wood anemone and Sheep Pasture get their names? What does the *queset* of Queset Brook signify in the language of Native Americans? Scratch a name in a landscape, and history bubbles up like a spring.

In my daily rambles along the path, I have been inspired by a famous observer of the Irish landscape, the early-twentieth-century naturalist Robert Lloyd Praeger, who walked over all of Ireland "with reverent feet," he said, eschewing motor transport, "stopping often, watching closely, listening carefully." And although I have aspired to Praeger's pedal reverence, I know I have fallen short. Another thirty-seven years walking my path would not do it justice. The contemporary writer and cartographer Tim Robinson, another close observer of the Irish landscape, defines something he calls the "adequate step," a step worthy of the landscape it traverses. The adequate step takes note of geology, biology, myths, history, and politics, says Robinson in *Stones of Aran*. It also includes the consciousness of the walker. And even all of that, he states, is not enough. No step, or series of steps, can ever be fully adequate. "To forget the dimensions of the step is to forgo our honor as human beings," he writes, "but an awareness of them equal to the involuted complexities under foot at any given moment would be a crushing backload to carry."

A crushing backload, indeed: fiddlehead ferns, downy woodpecker, pickerel, granite flake, Canada mayflower, moonrise, bluebirds, spring peepers, monarch butterflies, glacial scratches on bedrock, and, of course, the human history of my path, which in its transformations over the centuries encapsulates in many surprising ways the history of our nation and of our fickle love affair with the natural world. Step by step, year by year, the landscape I traversed became deeper, richer, more multidimensional, always overflowing the mind that sought to contain it. Ultimately, almost without my willing it, the path became more than a walk, more than an education, more than a life; it became *the* Path, a *Tao* (Way), a thread that ties one human life and the universe together.

A weed plucked at the side of the path might have found its way to the New World in a seventeenth-century sailing ship. Scratches on a rocky ledge evoke colossal mountain-building events on the other side of the world millions of years ago that modified the planet's climate and caused glaciers to creep across New England. The oxygen atoms I suck into my lungs were forged in stars that lived and died long before the Earth was born. It is something of a cliché to say that everything is connected to everything else, but when you know one place well—not just intellectually but with the deep-gut knowledge that enters through the soles of your feet—connections just keep popping up. A character in Anne Michaels's novel *Fugitive Pieces* says: "If you know one landscape well, you will look at all other landscapes differently. And if you learn to love one place, sometimes you can also learn to love another." Having learned to know and love my path in all of its local abundance, the light-years and the eons no longer seem quite so forbidding, tropical rain forests and droughty deserts seem not so far away. A minute lived attentively can contain a millennium; an adequate step can span the planet.

Measured from end to end, the path described in the following pages is hardly more than a mile, but the territory it traverses is as big as the universe. I don't mean to suggest that my path is in any way special; it is commonplace—about 3,000 paces of more or less typical New England landscape. It is the fact that it *is* commonplace that is the point of this book. Any path can become *the* Path if attended to with care, without preconceptions, informed by knowledge, and open to surprise.

VILLAGE

IN 1850 the great impresario P. T. Barnum brought to the United States a young coloratura soprano who had already captured the affections of Europe. Her irresistible name was Jenny Lind, and she became America's first pop star. For two years, she entranced audiences up and down the Eastern Seaboard with a voice that was said to be divine—although Barnum's hype may have been as much a part of Jenny's success as her talent. In the wake of her visit, and for decades afterward, schools, streets, public buildings, even towns were named for the Swedish Nightingale.

Swedish immigrants comprised a substantial part of the workforce at the Ames Shovel Company of North Easton, Massachusetts, twenty miles south of Boston. In the 1870s the company and the village were at the height of their prosperity; it is said that three-fifths of all the shovels manufactured in the world at that time were made in the Ames factories. New streets were laid out to accommodate the village's growing population. One of these was home to families named Swanson, Lonn, Anderson, and Lindquist. Not surprisingly, this street was christened for the most famous Swede of all.

My walk to work each day starts along Jenny Lind Street; it is the first part of the path. The houses on the street are typical late-nineteenth-century mill-town housing: two-story, front-gabled, timber-frame-and-clapboard construction. Many of the houses have modern additions at the sides or back, and perhaps a tacked-on porch, but it takes little imagination to see these structures as they must have looked in the 1870s and 1880s, as they went up one by one according to a standard plan in the mind of a local master carpenter. With my wife, I raised a family of four children in one of these houses, just around the corner from Jenny Lind Street. Even in the second half of the twentieth century it provided comfortable accommodation for the six of us. These sturdy homes must have seemed luxurious to the Ames factory workers who woke in the morning when the village wake-up bell was rung a half hour before dawn, who spent ten to twelve hours a day fabricating shovels at the factory, and who undoubtedly collapsed exhausted into bed not long after the curfew bell was rung at 9 P.M.

For all of the long working hours, nineteenth-century North Easton must have been close to the American immigrant's dream: a one-company village with neatly laid-out streets, uncrowded neighborhoods, leafy trees, handsome parks, and schools, public buildings, and churches donated and maintained by the town's preeminent family, the Ameses. Yankees, Irish, Swedes, and Portuguese lived and worked side by side in apparent harmony; within a few generations they would intermarry and contribute their progeny to the American melting pot. Even at the height of late-nineteenth-century labor unrest in other parts of the country, workers in the North Easton shovel works remained loyal to their company. And the Ameses remained loyal to the town; as the family's political influence and entrepreneurship extended far beyond the borders of the town (to include the governorship of Massachusetts and pushing the first transcontinental railroad across the United States), the village of North Easton remained the focus of their family life and public beneficence. Their gifts to the town are still evident. Three public buildings (and two private structures) were commissioned by the Ameses from the illustrious architect Henry Hobson Richardson, and the stamp of the famous landscape architect Frederick Law Olmsted is everywhere to be seen. The village is today a place of pilgrimage for students of architecture and landscape design.

When twenty-four-year-old Oliver Ames came to North Easton in 1803, the town already had a nascent industrial character, and Oliver was shrewd enough to see that the potential for development had only been lightly tapped. He had a product that a growing nation needed—it would take a lot of shovels to build the Erie Canal, then the Baltimore and Ohio Railroad. Later, the Civil War provided an apparently endless market for entrenching tools, and after the war the intercontinental railroad would be built with Ames shovels and Ames leadership. A great family fortune was acquired by forging steel and hardwood into tools for continental conquest, and samples of those shovels today reside in the Tofias Industrial Archives of Stonehill College. There is a shovel shaped for every purpose, and in their collectivity they constitute a kind of history of nineteenth-century America: canals, railroads, water power, mining, steam, agriculture, war, and, as the century matured, the first purely aesthetic works of landscape architecture, such as Olmsted's great city parks (in New York a workforce of 3,600 men moved millions of tons of earth to create Central Park). The Ames shovel was an instrument by which visionary men could shape the landscape to their dreams and desires, but only as much material could be moved at a time as a man could lift. The resulting works were ensured a human scale.

North Easton offered two natural resources which Oliver recognized, and which he and his sons brilliantly exploited: water and gravity. The water flowed in the Queset Brook, a relatively minor stream that one can almost jump across as it tumbles through North Easton village. The brook rises with trickles of springwater in the highlands north and west of town, then gathers into a modest but reliable flow as it passes near the center of the village. In the Queset's mile-long run through the industrial quarter of the village, it drops fifty feet. When a cubic foot of water drops through fifty feet, it surrenders enough energy to run a dozen modern washing machines for the time it takes the water to fall. Keep many cubic feet of water coming in a steady stream, and you have the power to hammer shovel blades from chunks of iron, to edge and polish blades, to turn hardwood logs into handles, to fix handles to blades.

Men with sledges could make shovels with muscle power alone, as they had done for centuries, but not as cheaply, reliably, or in such volume as was possible with water power. The babbling Queset Brook, in its carefree tumble from the resistant granite highlands north of town to the more deeply eroded sedimentary plains to the south, when properly harnessed was an ever-flowing source of wealth. Oliver Ames harnessed it and, with his sons, Oakes and Oliver II, turned water and gravity into a family fortune.

I LOVE THE subtle sounds of my morning walk along Jenny Lind Street. I listen to the amorous suit of a bright red cardinal in a tree by the Catholic church, and the who-gives-a-damn chatter of jays at backyard feeders. Through windows open just a crack, I hear the muffled music of kitchen radios, and mothers shouting to slug-a-bed kids. It would not have been so in the era of the shovel shops. The great water-powered hammers thundered all day long, from 7 A.M. to 6 P.M., presumably obliterating the sounds of nature and domestic life. It is said that the banging could be heard a mile or more away. Students in the grammar school and high school, on high ground in the middle of the village, studied their lessons and practiced their drills to the steady *thud, thud, thud* of the hammers. The Swedish housewives in their tidy frame houses on Jenny Lind Street performed their morning chores to the booming timpani of shovel making. Not even the Ameses were exempt from the din; the homes of the first two generations of the family were situated hard by the factories. The squeak of turning waterwheels, the bang of hammers, the rasping buzz of saws—this was the aural background of the Industrial Revolution in America. In North Easton, the driving force of this cacophony was the gentle Queset Brook.

Water was sacred in all ancient cultures—every pool and stream was believed to have its resident spirit. Against this overlay of sacred myth, the mundane idea that water could be used as a source of energy was hard won. The first record of a waterwheel harnessed to turn a millstone comes from the Roman world of the first century B.C.E., but the Romans

12

were not in a hurry to exploit waterpower, perhaps because it would throw so many human slaves out of work. Only when Constantine turned the empire toward Christianity in the fourth century C.E. did water mills become commonplace, perhaps because there were fewer slaves to do the work, or perhaps because monotheistic Christians were less troubled than pagans about resident sprites of pools and streams. In general, Christians had few prejudices against technology; after all, didn't God give Adam dominion over the planet and its creatures, commanding him to "subdue the Earth"? Christians may have had their eyes fixed on Heaven, but their feet were firmly planted on terra firma. In the days of Oliver Ames, Christians were not the least reluctant to ask flowing water to drive their mills; for the pious, patriarchal entrepreneurs of the Industrial Revolution, a turning waterwheel squeaked out its owner's prayer: *O Lord, increase my bounty that I may serve you.*

Not far from the end of the path, in a wooded part of the Stonehill College campus, is an eighteenth-century millstone quarry. Several unfinished millstones remain part of the living rock, rising up from the forest floor like Druidic altars. One large millstone, five feet in diameter, was finished by the quarriers but never moved. Some years ago a group of stalwart students tried to move the stone to the central campus quad, unsuccessfully. The millstone would have made a fitting centerpiece for the campus, which is a former Ames family estate. The stone, which I calculate to be about three tons, was almost certainly meant to be turned by the modest Queset Brook, a superb testimony to the power of water when gravity causes it to flow downhill.

What gravity is and why it is, no one knows. Albert Einstein spent most of his life trying to figure it out, but the secret eluded him. It is simply a fact that everything in the universe with mass pulls on everything else. If it weren't for the initial outward impetus of the Big Bang, gravity would have caused the entire universe to collapse into a heap. (Indeed, someday the cosmic collapse may happen, if and when the initial impetus is expended, although the best evidence suggests that the expansion will go on forever.) According to present theories, the universe began about 15 billion years ago as an explosion from an infinitely small, infinitely hot seed of pure energy. As the primeval fireball expanded and cooled, matter—hydrogen and helium—condensed from radiation. The first stars and galaxies were born as gravity pulled the hydrogen and helium gas together into massive spheres and eddies. As stars burn, they fuse hydrogen nuclei into helium and then into heavier elements, and when stars die explosively, they scatter those heavy elements into space. After many generations of exploding stars had seeded the universe with carbon, oxygen, silicon, and iron, our own Earth and Sun were squeezed into existence by gravity.

When atomic nuclei fuse at the core of a star, some of their mass is turned into pure energy, according to Einstein's famous equation, $E = mc^2$ (energy equals mass times the speed of light squared). Every second at the Sun's center, 660 million tons of hydrogen are fused into 655 million tons of helium, and the missing 5 million tons of matter ultimately appears at the Sun's surface as heat and light radiated into space. A tiny fraction of the Sun's energy falls upon the Earth's oceans and evaporates water molecules into the air. It takes about 1,000 calories of energy for the Sun to evaporate a thimbleful of water from the sea; each thimbleful of water in the atmosphere represents 1,000 calories of stored solar energy. The Sun does the heavy lifting on Earth, heaving tens of thousands of cubic miles of water up out of the seas and into the atmosphere each year. Most of this water precipitates back into the oceans, but some of it falls on land as rain or snow, from whence it makes its way downhill to the sea in a great recirculation called the water cycle. Around and around the water has cycled for 4 billion years, since the oceans were born, in trickle or torrent, eroding and shaping the land, bringing fresh water to land plants and animals, providing terrestrial habitats for life.

North Easton's Queset Brook is a tiny conduit in this unceasing flow. Young Oliver Ames looked at the Queset Brook and saw God's solar-powered engine, providential energy waiting to be tapped. He built dams to the north and west of the village to impound water during times when the factory was idle or when rain swelled the streams beyond the capacity of the mills to use the excess. Water behind a dam is like money in the bank. It can be made to turn a wheel; it can pound and grind and drill.

Today, North Easton's shovel shops are silent, and the Sun-lifted energy of the Queset Brook is expended as a pleasing babble only, as the stream tumbles untroubled from pond to pond. Standing beside the brook, one finds it hard to imagine how these placid waters ever supplied the force that made the factories go. But occasionally the babbling thimblefuls of water, when gathered in prodigious numbers, do shout and roar. This happened on February 12, 1886, when heavy rain on frozen ground filled the Long Pond Dam north of the village to overflowing. Ames Company agents worked heroically to shore up the dam against the gathering flood, but eventually water breached the barrier and rushed downstream through the village, washing out a section of railway track. More serious damage was narrowly averted.

Not so in the Great Flood of March 18, 1968, when there were no company personnel around to mind the dams or tend repairs. All morning and afternoon the sky gushed water, six inches of rain in a few hours. At 5:30 P.M., without warning, the dam at Flyaway Pond west of town gave way with a sudden roar, and 350,000 tons of water were released onto the village. Trees and automobiles were swept away.

Houses were battered. Massive stone walls tumbled. It was a sudden, vivid indication of the Sun-stored energy that a century earlier had banged out the majority of the world's shovels.

WHEN HENRY DAVID THOREAU and his brother John took their famous boating trip along the Concord and Merrimack Rivers in the summer of 1839, they too witnessed a demonstration of the power of water and gravity. The Concord River takes a more or less level course through flat meadows; its reedy banks were then (and remain today) home to a rich proliferation of wildlife. The Merrimack River, by contrast, tumbles downhill out of the granite mountains of New Hampshire, and in Thoreau's time its banks were lined with working mills, replacing the "wild musical sounds" of the Concord with the din of industry. As they camped along the Merrimack, even on Sunday, Henry and John were kept awake at night by the raucous sport of mill laborers, who were "unwearied and unresting on this seventh day." Similarly, anyone camped along North Easton's Queset Brook in that same year would have heard the hammers pounding all day long—a far cry from the birdsong and insect buzz that hymned the Queset Brook before the coming of Oliver Ames, and which hymn it again now that the mills are gone.

The Queset is a small stream; it hasn't nearly the

volume of Rhode Island's Blackstone River or New Hampshire's Merrimack River, those wider torrents that turned so many wheels of America's Industrial Revolution. The Queset's capacity for work was soon found wanting. In 1821 the Ames factories were producing 370 dozen shovels a year, and this output taxed the available water supply. Upstream of the village, dams were thrown across every tributary of the Queset, catching every available drop of water. Production rose to 4,000 dozen shovels in 1832, and 20,000 dozen shovels in 1844. Eventually, the power of the brook was maxed out, all those thimblefuls of solar energy squeezed of their power. If production was to expand further, new sources of energy had to be found. In 1852 the company installed the first steam engine, and the village began to sever its economic relationship with Queset Brook.

STEAM POWER IS no less a gift of the Sun than tumbling water, and to understand this story, too, one must go back to the beginning. When the first single-celled organisms appeared on Earth, more than 3 billion years ago, they fed upon carbohydrates—sugars—dissolved in the sea. The sugars had their origin in chance chemical reactions. Life, however, multiplied exponentially; one cell made two, two made four, four made eight, and so on. Self-replication is the essence of life. It was inevitable that burgeoning organisms in the sea would outstrip their catch-as-catch-can food supply. It would seem that life was doomed to a dead end, but before the sugars ran out, certain organisms evolved the ability to use the energy of sunlight to synthesize carbohydrates from carbon dioxide and water, a chemical reaction called photosynthesis, catalyzed by chlorophyll. And so the first plants appeared on Earth, manufacturing their own fuel. No longer did those earliest organisms live a hunter-gatherer existence, scrounging fuel from the sea; now they settled down and became farmers, so to speak, making their own food. And the cells that could not do photosynthesis fed upon cells that could. The planet greened, first with photosynthesizing algae in the sea, later with multicelled land plants.

Coal is fossilized plants. The coal seams of North America were once swampy forests that covered parts of the Earth's surface hundreds of millions of years ago. The trees, with their energy-packed carbohydrates, died, fell to the forest floor, and were eventually buried and compressed into coal. Like a thimbleful of water lifted into the atmosphere by evaporation, a lump of coal is a packet of stored sunlight. When the water of the Queset Brook could no longer supply sufficient energy to drive their shops, the Ameses turned to energy buried within 300-million-year-old forests. When the first steam engines were installed in the Ames factories in the early 1850s, they were fed with Nova Scotia coal, schoonered to Fall River along the coast, then hauled by ox-drawn wagon or sleigh to North Easton. In 1855 the first railroad reached the village with coal from Pennsylvania—and there was no turning back.

A steam engine is a miniature version of the planet's water cycle. Wood or coal is burned to heat liquid water into vapor. The hot vapor at high pressure pushes a piston in an iron cylinder. At the end of the stroke the steam is condensed back to liquid water, and the resulting partial vacuum in the cylinder allows atmospheric pressure to drive the piston back down the cylinder. This linear stroke, back and forth, is converted to rotary motion by a system of levers. With the steam engine, industrialists had an efficient and reliable source of rotary power to drive their factories. But coal-fired energy was not as cheap as waterpower, and until the Queset Brook had given up its last babble of gravitational energy the Ameses had no motive to switch to steam.

During the second half of the nineteenth century, when the Queset Brook could no longer supply the requisite energy, North Easton schoolchildren did their lessons to the *chug* and *hiss* of iron engines. The waterwheels were abandoned, and the superfluous brook became no more than a town amenity, the ponds behind the dams pleasant places of recreation. But the location of the factories was all wrong; they were far removed from the source of coal, and even more distant from the primary markets for shovels in the developing West. Early in the twentieth century the company purchased manufacturing facilities in Parkersburg, West Virginia. In 1932 the headquarters of the Ames Shovel Company followed the forges and hammers to coal-rich Appalachia. Exactly 100 years after the introduction of the first steam engine in North Easton, the last shovel-manufacturing facility in town began closing down.

THE SHOVEL SHOPS are gone, and the houses along Jenny Lind Street are no longer inhabited by Swedes. The street is rigged with electric lines, and wires for telephone, cable television, and broadband Internet access. Many of the carbon-copy two-story houses from the 1870s and 1880s have been doubled in size by modern additions, and in the driveways sit a car or two (or three). The original clapboards as often as not are covered with vinyl siding. The central village, too, has changed. Shopping malls in the outer parts of town have replaced Tom Barnhill's five-and-dime and Charlie Harvey's food market. The old Ames shovel shops have been converted to nonmanufacturing uses: light warehousing, insurance offices, even a health club. Locomotives no longer huff and puff through town, delivering coal to the factory, taking out shovels. The tracks are overgrown, the bridge over the tracks at Main Street replaced with fill, the switches that once shunted cars from main line to factory siding immobilized by rust.

North Easton is now a desirable bedroom community for Boston, and the descendants of Irish, Portuguese, and Swedish factory workers go to school with recent arrivals from every part of the country, including Blacks, Asians, and Hispanics. The schools endowed by Ames largesse, prominent on their hilltop locations at the village center, have been closed and replaced by more modern facilities on the village outskirts; education, once prominently on display, is now tucked out of sight. According to census lists for Jenny Lind Street, in 1916 the homeowners (mostly Swedes) included a tailor, shoemakers, a stonemason, carpenters, bakers, a coachman, a clerk, a gardener, a plumber, a milk dealer, and, of course, shovel makers. The 2000 census lists teachers, retail clerks, a systems analyst, a claims adjustor, a dental hygienist, a psychologist, a nuclear medicine technician, a travel agency manager, an attorney, a real estate broker, and—not to be found on the 1916 list—retirees. The transformation is striking; manual labor has given way to services, material products to ideas. Working hours are shorter, life spans longer. There is probably not a person among the current residents of Jenny Lind Street who hankers for the Ames factory wake-up bell that rang a half hour before dawn, the all-day bang of the hammers, the long work hours, and paucity of leisure. Evolution doesn't sit still. The arrow of history was shot from nature's bow 15 billion years ago and won't be put back.

One hundred and fifty years ago the inhabitants of Jenny Lind Street walked to work, streaming with other villagers to the factory floors from homes within sound of the coal-fired engines. Today, one by one, they climb into their cars and drive off toward Boston, as likely as not to one of the hundreds of computer firms strung out along Boston's Route 128, "America's Technology Highway." The fossil fuel that previously turned out shovels now powers the hundreds of automobiles that leave the village each morning to creep in ever-thickening traffic jams toward faraway places of work. Collective village life as it was known in the nineteenth-century has more or less come to an end. It doesn't really matter anymore where one lives; the automobile and school bus take us where we want to go. As I write, the citizens of Easton are vigorously opposing the return of the railroad to the town, this time to carry commuters to and from Boston. The private car reigns supreme, and the lovely railroad station designed by Henry Hobson Richardson is now headquarters of the Easton Historical Society.

SOMETHING PROFOUND happened to America when the Ameses and other nineteenth-century entrepreneurs switched from waterpower to steam, and later to internal combustion. Water—a clean, renewable resource—was replaced by fossil fuels. Worries about greenhouse warming, the exploitation of remaining wilderness areas for coal, oil, and gas, and the inexorable spread of asphalt paving now have caused many people to hanker back to early-nineteenth-century village life, which was clustered near the factories and tumbling streams. Thoreau might have balked at the noise and clamor of North Easton had he visited in the age of water-power, but surely the village was a realization of mythic Arcadia, that imagined time and place in the Greek Peloponnese when spiritual and economic values were in happy balance, a tamed and congenial countryside beyond the urban hubbub where people lived in proximity to wild nature and earned their bread within reach of shank's mare. The Arcadian character of the village was nurtured by the beneficence of the Ames family and the talents of men like Richardson and Olmsted, whose work was imbued with the nature-worshiping sensitivities of Thoreau and Emerson. As Peter Schmitt has shown in his *Back to Nature: The Arcadian Myth in Urban America*, the dream of combining urban enterprise with wilderness sensibilities became particularly strong in the late nineteenth century, at precisely the time when the Ames estates of North Easton were taking shape, including the one that contains my path as I enter the woods at the end of Jenny Lind Street.

But however one chooses to romanticize what in retrospect seems a fetching life, it is impossible to reclaim it. Damming every stream in America will not slake our need for energy. Fossil fuels supply that need for the time being; they also confer a freedom and mobility that most Americans seem to prize. Technology—with its awesome potential and perils—is here to stay.

WOODS

JENNY LIND STREET dead-ends at Seaver Street. As the town's population increased in the late nineteenth century, it would have been logical to extend Jenny Lind across Seaver, away from the village center, making more sites for the standard-issue, timber-frame, front-gable houses that characterize the older parts of the village. But before that happened, a third generation of Ameses began buying up land for beautifully landscaped and magnificently mansioned estates that ultimately embraced the village and halted its growth in every direction but one. From the end of Jenny Lind, my path makes a short jog along Seaver, then enters woods that were once part of an Ames estate called Sheep Pasture, and which are now in the care of the Natural Resources Trust of Easton.

This is a scruffy woodlands, mostly oaks and white pine, with some maple and hickory. No Ames-employed woodsman clears the understory today, as he might have done a hundred years ago, so my progress is confined to a well-worn track, three or four feet wide, that is kept open by villagers who walk their dogs, jog, take the air, or, in winter, cross-country ski. Beside the path, sunlight falls onto a narrow subcanopy of sassafras, dogwood, or occasional cherry. My favorite time for walking the wooded path is spring, when the ground under the trees is carpeted with Canada mayflower, also called wild-lily-of-the-valley. The plant insinuates a web of runners beneath the leaf litter, colonizing dark continents of the forest floor. From this hidden infrastructure of communication and transport, it throws up thousands of paired green leaves, one to the right and one to the left on each stem, like supplicant hands. In a few weeks time there will be tiny white flowers and the plant will be busy with arrangements for the next generation, but for the moment the business is pure energy. Capturing sunlight. Soaking up the rays.

There is little enough sunlight on the woodland floor; the Canada mayflower aggressively courts its share. A phrase from a poem by Andrew Marvell aptly describes the mayflower: "a green Thought in a green Shade." Each small plant would seem to have one thought in mind: Catch whatever light it can before the deciduous trees complete their leafy canopy. When you think about it, there is no need for plants to rise high on stems or trunks; there would be just as much sunlight to go around if plants spread themselves out on the ground and shared and shared alike. But in nature, competition, not sharing, is the rule. Plants evolved tall stems and trunks to put their neighbors in the shade. Oaks, maples, and hickories stretch skyward, putting out buds, doing their best to leave the Canada mayflowers in the lurch. But mayflowers manage on whatever dollops of light slip through the canopy. They get a jump on deciduous trees by blooming early.

No law of physics is more basic than the law of entropy, the tendency of the universe to move toward disorder and death. But life bucks the tide, using available free energy wherever it can get it, and hereabouts the most abundant source of energy is sunlight. The mayflower constructs its tiny oasis of order by drawing upon a corresponding increase of disorder at the center of the Sun, where hydrogen is fused into helium. There, deep at the heart of our planet's star, is the source of the energy that drove the Ames shovel shops and drives the flower too. The Ameses turned radiant solar energy into a family fortune, by harnessing waterpower and the energy of coal; the Canada mayflower does it too, amassing its own Midas wealth of carbohydrates. In summer, about a hundred-millionth of an ounce of the Sun's depleted mass (multiplied by the speed of light squared) falls each second onto these woodlands; in winter less than half as much. A fraction of a millionth of an ounce of matter turned into energy is all it takes to tip the balance of the season from winter toward summer. A fraction of a millionth of an ounce of fused hydrogen is all it takes to rocket the Canada mayflower up out of the ground. The leaves of the plant push aside the detritus of ruinous winter, thrusting toward light, its fuse lit by a match scratched at the Sun's core. Each pair of green leaves soaks up sunlight, using the radiant energy to make sugar through the magic of photosynthesis.

Biologists are not sure how photosynthesis evolved, although they have likely scenarios; that it happened early in the history of life is certain. Every high school student learns the basic equation: Carbon dioxide plus water plus sunlight yields sugar and oxygen (with the Cs, Hs, and Os appropriately balanced). The equation doesn't nearly convey the complex chemical reactions that connect one side of the reaction arrow to the other. Crucial to the process is a boxy molecule with a magnesium-nitrogen heart and a long carbon-hydrogen tail: chlorophyll. Atomic electrons in the molecule absorb solar photons and are bumped up in energy. As they return their bounty, they energize reactions that create intermediate products called ATP and NADPH, which then move along the assembly line. When all is said and done, it is sugar that appears at the factory door. I love to think of all this molecular activity going on in the splayed leaves of the Canada mayflower. The only hint of this humming manufactory is the color of the leaves.

Chlorophyll absorbs energy from the blue and red parts of the solar spectrum—the ends of the rainbow. The green middle of the spectrum is reflected, and that's the light that enters our eyes. Green is not the color of photosynthesis, as one might assume. Rather, green is the leftovers of the solar feast. If the absorption of sunlight were more efficient, and the entire solar spectrum were used by the plant, the leaves of plants would be a nonreflecting black. It's as if nature limited its consumption, gobbling the bulk of sugar-building light but generously tossing us scraps of green.

Green plants give us more than beauty. As animals, we burn sugar when we respire; that's where we get our pep. Every breath of oxygen sets little fires alight in our lungs. Burning sugar keeps us warm and active. But we can't make sugar ourselves; for that we need plants. We need to eat green plants, or the flesh of animals that eat green plants. The products of photosynthesis are as vital to us as air and water. Every animal on Earth needs its portion of the green plants' fortune, although humans get the lion's share. It is estimated that our species currently commands between one-third and one-half of all products of terrestrial photosynthesis, as food for ourselves and our domesticated animals, but also for fuel, building material, and fiber. Worldwide, humans are harvesting (or needlessly destroying) trees at an alarming rate. Satellite surveys show a total area of tropical forest the size of West Virginia being cleared annually. The long-term consequences of such rampant deforestation are unclear, but it is known that plants maintain the balance of carbon dioxide in the atmosphere and carbon dioxide traps heat from the Sun (the so-called greenhouse effect). If the total mass of green plants is substantially reduced, the level of carbon dioxide will rise. The wholesale destruction of tropical forests might therefore contribute to warming climate, melting ice, and rising sea levels.

But it is the effect of deforestation on biodiversity that most immediately concerns biologists. Tropical forests are habitats for an astonishingly rich variety of plants and animals. If the present rate of clear-cutting continues, all tropical forests may be gone before the end of the century, and with them possibly 50 percent of all species of life on Earth. No comparable biological catastrophe has occurred on this planet since the time of the dinosaur extinctions 64 million years ago, when the impact of a massive asteroid blasted as many as half of all living species into oblivion.

Global warming and extinction doomsday scenarios may not come to pass, if for no other reason than that humans have the technological ability to create as well as destroy. Here in New England, the woodlands have made an impressive comeback. There are more forested acres in my town today than at any time since the seventeenth century. The high point of deforestation coincided with the heyday of shovel making. Then every tree was a valuable industrial resource. Two acres of forest converted into charcoal were required to produce a ton of iron, still more to forge the iron into shovel blades. Timber was also required for shovel handles, to build the shops where the shovels were made, and to house the workers. No one worried then about deforestation. And even before the charcoal-hungry furnaces and the neat rows of timber houses, the land had been cleared for agriculture. An old map of Easton, dated 1830, shows pastures and meadows everywhere. From anywhere along my presently wooded path I might then have looked across open fields to the steeple of the village church and the belfry where the Ames factory bell tolled its call to work.

When the first English settlers came to New England, their most characteristic implement was the ax. They hewed English farms out of an old-growth wilderness of hemlock, beech, oak, birch, and pine. Native Americans also cleared land for agriculture—small fields near settlements, mostly by burning—but only as necessary. When nutrients in the soil were exhausted, the community moved on to another place (land was held in common), and fields reverted to forest. By contrast, the English cleared with abandon and staked private claim to the land. Trees were the most obvious and accessible resource of the New World. The woods rang loud with the strokes of colonial axes. Sawmills buzzed with the rasp of iron. Back in the old country, woodlands had been managed for centuries as a sustainable resource for building material and fuel; scarcity was the motive for old-world conservation, and even today, early in the twenty-first century, English woodsmen widely practice *coppicing* and *pollarding,* renewable techniques of harvesting timber by pruning low shoots and high branches, respectively, without killing the trees. But the vast forests of colonial America presented the appearance of inexhaustibility. Any conservation ethic brought to the New World by the colonists was soon abandoned. The Pilgrims stepped ashore at Plymouth Rock with axes swinging, and forests fell before their onslaught.

In 1621, one year after the foundation of the Plymouth Colony, the first vessel to return to the mother country with products of the New World carried two barrels of furs; the remainder of the cargo was clapboards, "as full as she could stow." As the New England colonies grew, farmers cleared pastures for their burgeoning numbers of cattle, sheep, and swine. They required firewood for winter warmth, boards for building, charcoal for forging. No half-timbered houses satisfied the "new" Englanders; their homes were fully timbered. Wooden roofing shingles replaced the thatch and slate of the old country. Rail fences substituted for the living hedgerows the settlers had known in Europe. Within a century, the greater part of southern

New England's forests were gone, replaced by a patchwork of fields and homesteads.

Still, New England was not an ideal farming environment. There was nothing wrong with the fertility of the soil; there was just so little of it. Once the trees were gone and the thin layer of humus eroded, what the colonists discovered underfoot was the rocky detritus of glaciers. It was not easy coaxing crops out of this hardscrabble land. So at the beginning of the nineteenth century, when canals and railroads opened up less stony land in western New York and Ohio, New England farmers packed up their animals and children, their butter churns and spinning wheels, their axes and hoes, and headed west, worn out by the incessant task of heaving boulders from in front of their plows. When they departed, their abandoned houses collapsed into cellar holes and rotted. Wells became filled. For a while, the hungry furnaces of the factories and the building of homes for an immigrant tide continued to consume timber. But eventually the noise of iron blades against the trunks of trees went silent, and the forest began to reclaim the land. As I walk the wooded path today, the threatened rain forests of Brazil, Zaire, Malaysia, or Sri Lanka seem far away. It is easy to forget the thousands of imperiled plants and animals of the tropics when our own gloriously resurgent New England woodlands are burgeoning with Canada mayflowers. That woodland carpet of irrepressible green is reason enough to incline one toward optimism, even in a world where the destruction of greenery occurs on every side.

WHEN OLIVER AMES came to North Easton in 1803, he had only two things on his mind: making a perfect shovel and a comfortable living. His sons, Oakes and Oliver II, who took over management of the company in 1844, had in mind making the most economically lucrative shovel and a more than comfortable living; they envisioned great wealth as their natural right and believed the resources of the continent to be God's providential gift to those who had the vision and determination to seize them. It was a time in America of indiscriminate and profligate exploitation of nature's bounty: trees, coal, bison, passenger pigeons, even human labor. Once the barrier of the Appalachians had been breached, in no small part with the help of Ames shovels, the continent unrolled to the west like an inexhaustible smorgasbord of plants, animals, and minerals. With the encouragement of their friend Abraham Lincoln (Oakes served in Congress from 1862 until his death in 1873 as a representative from Massachusetts), the brothers undertook to drive the Union Pacific Railroad across the continent, and Oakes and Oliver II as much as any other persons can be credited with the success of that dramatic undertaking. Given the tenor of their time, they could hardly avoid being touched by the American ethic of extravagant squander, although their factory village of North Easton was about as close as one might find to the Industrial Revolution working at its best. Oakes's and Oliver II's homes in the village were certainly ample, but ornament and opulence were not high on their agenda. I suspect that when the brothers looked at a tree, perhaps even a lovely elm outside a dining-room window, what they saw was a dozen shovel handles or a bushel of charcoal.

As is so often the case in families built on industrial fortunes, it was the third generation of Ameses who "got culture," who ornamented the town with the designs of the greatest architects and artists of their time, and who patronized the arts and sciences. It was this generation who could look at a tree and see a public ornament, a thing with a fit and important place in God's great scheme of things. The family established a fund, still existing, for planting trees along the town's highways and byways. On the top floor of the town's Ames Free Library are fourteen huge volumes that were among the library's first acquisitions, and still, a hundred years later, remain the largest volumes in the collection: Charles Sprague Sargent's *Silva of North America*. A silva is a description of the trees of a certain area. Sargent undertook to describe all the trees of the North American continent. His fourteen-volume compendium is one of the great works of nineteenth-century science, and the third generation of Ameses, relieved by great wealth from the pressures of making a living, were among his patrons and admirers.

The *Silva of North America* is as Yankee as cod and as Boston as baked beans, exactly suited to the taste of a generation of prosperous, proper New Englanders who no longer needed to exploit the forests. The days of the region's diminishing woodlands were past. A new generation of businessmen, with taste and affluence, could begin rebuilding the landscape from the ground up, with proper scientific and artistic sensitivity. Charles Sprague Sargent and his great botanical project gave their dreams a veneer of intellectual propriety. After all, Sargent was one of their own, born into one of the great Yankee families of Boston founded on commerce. The Sargent genealogical tree includes prominent Bostonian names like Saltonstall, Brooks, Winthrop, Everett, Gray, Ward, and Hunnewell. The family is perhaps best known for the painter John Singer Sargent and Governor Francis W. Sargent, but many other Sargents distinguished themselves in business, public service, or the arts. Charles Sprague's father was a successful businessman. When Charles was born, the family lived on Joy Street on Beacon Hill. Soon they moved permanently to their summer estate in Brookline, called Holm Lea, not all that far from downtown. As close as it was to the center of Boston's commercial and cultural life, Holm Lea consisted of

130 acres of handsome parkland and gardens. It was the largest personal estate so close to Boston.

Charles had a classic Yankee education—private school, then Harvard. He served in the Union army during the Civil War and traveled in Europe. When he returned to Boston in 1868, his record as a scholar, soldier, and traveler gave no hint of future distinction. Twenty-seven years old and disinclined to enter the family business, he took up the management of his father's estate and fell willy-nilly into horticulture. An interest in horticulture and the design of gracious garden estates was one of the common enthusiasms of moneyed gentlemen in Sargent's social class, among them the third and fourth generation Ameses of North Easton. In 1872, to everyone's surprise, President Charles Eliot of Harvard named Sargent professor of horticulture. The next year Sargent was given responsibility for Harvard's Botanic Garden in Cambridge and the new Arnold Arboretum in Jamaica Plain. According to his biographer, S. B. Sutton, Sargent's qualifications were limited—he "was little more than a glorified gardener." Of course, it helped to have the right connections, at a time and in a city where connections were everything.

However, Sargent soon confirmed Eliot's confidence. In addition to organizing one of the country's great arboretums, he began working on the *Silva*: writing the text, supervising the production of the magnificent illustrations by Charles Faxon, and traveling about the continent to collect specimens and observe trees in their natural habitats. When the work was published, it met with universal acclaim. The *Silva* is typically Victorian, a science of description and classification that prided itself on comprehensive collections proudly displayed. It was not yet ecology, and not yet a secure basis for conservation, but until nature's bounty was properly cataloged no further science was possible. Sargent devoted himself to that end.

To leaf through the fourteen huge volumes of Sargent's remarkable book is to take a journey back in time. Today's science tends to be abstract and cerebral; Sargent's *Silva* provides a good rush to the senses. The heft of the volumes, the big bold print, the lush Latin names, the anecdotal footnotes, the magisterial text, and above all the finely rendered illustrations of Charles Faxon appeal as much to the hand and eye as to the intellect. The *Silva of North America* is a kind of Arnold Arboretum of the printed page—a wealthy man's personal estate turned into a great instrument of public instruction. The naturalist John Muir wrote of Sargent, "While all his surroundings were drawing him toward a life of fine pleasure and the cultivation of the family fortune, he chose to live laborious days in God's forests, studying, cultivating, the whole continent as his garden."

The effusiveness of Muir's language is very un-Sargent, very un-Yankee, but the sentiment is exact. Charles Sprague Sargent was the archetypal Boston

Brahmin: aristocratic, aloof, taciturn, reserved. On the outside he was a bit of a cold fish, but like many Yankees, once fired with an inner passion, he had unflagging energy. He built the Arnold Arboretum into the magnificent institution it is today. He was an early champion in the cause of conservation and the creation of the national forests. He was friend of landscape architects Frederick Law Olmsted and Charles Eliot. His great book was a monument of its time, and an inspiration to other men and women of wealth across the country who created great parklands both public and private.

The natural outcome of Sargent's influence was realized in the fourth generation of the Ames family, in the persons of Oakes Ames and his wife, Blanche, Harvard botanist and botanical illustrator respectively, who early in the twentieth century built the most extensive of the Ames estates at Borderland, a few miles west of North Easton village, now a thickly forested state park. Appropriately, Oakes succeeded Charles Sprague Sargent as director of the Arnold Arboretum.

THE CANADA MAYFLOWERS on the woodland floor are followed in June by a spectacular wild orchid, the pink lady's slipper. I can never resist diverting myself from the path to inspect a clutch of lady's slippers beckoning from under the trees. I kneel, and slip my finger between the voluptuous folds of a blossom, a suggestive but inevitable reflex. Blanche Ames, mistress of Borderland, is widely admired for her paintings of tropical orchids. She would surely have fallen to her knees in the presence of the New England lady's slippers; what more appropriate posture for a botanical illustrator than that of prayer.

I have among my books a wildflower guide published in 1917. "It is becoming rarer every year," says this guide of the blossom, "until the finding of one in the deep forest, where it must now hide, has become the event of a day's walk." By midcentury the plant was so rare in our area that it was considered an endangered species. Now, with careful regulation and public education, the lady's slipper is making a comeback; in some parts of our woodlands the big pink blossoms droop their heads in crowded throngs. It is the rare villager today who would pick one of these flowers, although children playing in the woods might occasionally bring a bouquet home to Mother.

The woods of the path, with their scatter rugs of mayflowers and wild orchids, have a primeval feel about them; when the trees are in full leaf, it is easy to imagine that one is James Fenimore Cooper's Natty Bumppo tramping a virgin forest, though one is never more than a few hundred yards from the village. But, in fact, these woods are as much a human artifact as are the houses and automobiles I left behind on Jenny Lind Street. Almost nothing remains

here of the hardwood and conifer forests that re-
claimed the frozen tundra as glaciers retreated from
New England 12,000 years ago. Even those post–Ice
Age woodlands did not long exist untouched by
human design. As the ice began its northward retreat,
Asian hunter-gatherers trekked across the exposed
floor of the Bering Strait into Alaska (sea level was
hundreds of feet lower with so much ice on the land)
and then made their way through ice-walled passages
into the bountiful continent south of the glaciers. Almost immediately they imposed the imprint of
human technology upon the flora and fauna.

Many animals that had thrived in North America
during the Ice Age—including woolly mammoths,
mastodons, giant ground sloths, saber-toothed tigers,
and camels—became extinct, very likely because of
the pressure of human hunting. The forests, too,
were modified. The first Americans made clearings
for villages and fields. Vast tracts of forest understory
were burned to improve hunting. When European
settlers arrived, they cleared whatever forest was left
on halfway decent land. When Oliver Ames came to
Easton to start his shovel business, 80 percent or
more of southern New England's woodlands had already succumbed to fire or ax. The choicest trees
were sawed into planks or boards for house construction, or hewed for posts and beams. Some trees
were used for heating, fencing, or charcoal production. The rest were burned in the fields where they
fell, contributing to the fertility of the soil.

By the time the third and culturally astute generation of Ameses came along, farms were being abandoned for more fertile land to the west, and the
family was able to buy up tracts of land for the great
estates that would ring the town. They hired architect Henry Hobson Richardson to turn these
patchworks of small holdings into artful domiciles.
Richardson had recently dazzled his contemporaries
with his design for Trinity Church in Boston's Cop-
ley Square, and the Ameses were among the first to
recognize his genius. Richardson in turn employed
such gifted artisans as landscape architect Frederick
Law Olmsted, already famed for New York's Central
Park, architect Stanford White, stained-glass maker
John LaFarge, and sculptor Augustus St. Gaudens.
Seldom has a single village benefited from such an
assembly of talent. Richardson cultivated a style of
architecture that seemed to spring from the landscape itself, rather than from the spreading artifacts of
industrialization. He worked with native stone, even
to the extent of sometimes using undressed glacial
boulders. His goal in remaking the village of North
Easton was to enhance nature, rather than erase it.

The woods along my path, though now somewhat
unkempt, were a dream in Frederick Law Olmsted's
eye as he laid out the Sheep Pasture estate for Oliver
Ames, great-grandson of the founder of the shovel
factory, in the 1890s. These sculpted acres of mixed
hardwoods and pines were designed to be viewed
from across the cleared, gently sloping valley of the

Queset Brook from the veranda of the Sheep Pasture mansion. (Supposedly sheep once grazed on part of this land; hence the name.) The trees screened the village beyond, creating the pleasing prospect of a genteel English countryside. If the path through these woods and bordering meadows is today a source of visual delight, it is not so much nature's doing as Olmsted's. He applied the same principles of landscape design in such places as the Biltmore estate in Asheville, North Carolina, the grounds of the U.S. Capitol, the campus of Stanford University, Belle Isle in Detroit, Mount Royal in Montreal, and the parks systems of almost every major city in the United States. With Charles Sprague Sargent, forester Gifford Pinchot, conservationist Theodore Roosevelt, nature writer John Muir, and many other late-nineteenth- and early-twentieth-century environmentalists, Olmsted recognized the importance of managing the nation's natural resources and saving the most important natural amenities from uncontrolled exploitation. It was not untrammeled wilderness these men sought to save, but a tamed wildness, what landscape architect Andrew Jackson Downing called "a more refined kind of nature," a human artifact, a spiritually uplifting Arcadia.

WHEN I WALK the path at dawn, I am usually alone, except for the occasional person walking a dog. It is a measure of the village's postindustrial silence that the most characteristic sound of my early morning walks is the *tunk-tunk* of downy woodpeckers and the fainter, barely audible *tip-tip* of nuthatches. Sometimes I leave the path in hopes of catching a glimpse of a nuthatch clinging to the bark of a tree, upside down; imagine living your life standing on your head. Gray and red squirrels make hardly a sound as they scurry in the canopy, but signal their presence when a branch swishes back into place after an arboreal passage. Squirrels can make a racket when they want to, but they seem to respect the quiet of the early hour. Gray squirrels are everywhere in town, including the eaves of my house. The more dapper red squirrels have become rarer every year, until there is now only one part of our woods where I reliably find them. I don't know the reason for their decline; certainly they have fewer predators since boys gave up BB guns for video games. Perhaps the reason is competition from the more aggressive gray squirrels, who can be terrible bullies, not only to red squirrels but to birds at feeders too.

The inventory of Earth's living species currently stands somewhere near 2 million. There are almost certainly at least ten times as many species that have not yet been described and named—the true number of species may be more than 100 million. Many of these are inevitably doomed by human population growth. Wolves, bears, and rattlesnakes were once the objects of deliberate eradication campaigns in the town of Easton. The Reverend William Chaffin, nineteenth-century historian of our town, says that the last bear was killed in the mid–eighteenth century, and that a certain Benjamin Drake, an early resident, was paid five shillings in 1724 for ridding the community of wildcats. The last wolf disappeared at about the same time as bears. Laws were enacted in the eighteenth century to protect the few remaining deer, but by the time of Chaffin's writing in the late nineteenth century the deer were gone. Of course, deer are back now in uncomfortable numbers, destroying ornamental shrubs and gardens and colliding with cars, having learned to live, like the gray squirrels, in a kind of commensal relationship with humans. Beavers and coyotes have also returned, having formerly been harried into remote fastnesses. Bobolinks and meadowlark thrived hereabouts at the height of forest clearance, but shrank to a fraction of their former numbers as the woodlands returned. Today we coax them back from the brink of extinction because they have a particular hold on human affections. The pink lady's slipper, too, enchants our forests because we made the conscious decision to assist its return.

Contemporary conservationists rail against the wholesale clearing of forests in developing countries, and rightly so. But it is worthwhile remembering that something similar happened in New England hundreds of years ago: a pell-mell clearance of the land, a disruption of native habitats, and the introduction of many alien species. It happened, too, when the first Americans trekked south of the glaciers at the end of the Ice Age. It is inevitable that forest clearance and a diminishment of biodiversity will accompany third-world economic development, although what happened in New England—clearance by small-holding farmers with ax and saw—is a far cry from the use of the full panoply of twenty-first-century clearance technology. If conservationists from the developed world have anything important to give the peoples of developing nations, it is not self-righteous scoldings, or futile agitation to "save the wilderness," but an ethic of land and resource management such as that which has been so successfully applied along the path—a landscape lovingly contrived by the greatest landscape architect of the nineteenth century, now carefully managed by the Natural Resources Trust of Easton for the benefit of all the town's citizens. Ironically, this ethic of aesthetic land management had its origin to a large extent among the third- and fourth-generation offspring of the Industrial Revolution.

IF THE EARTH were not tipped on its axis, there would be no seasons. Climates, certainly—warm equatorial regions, temperate midlatitudes, frigid poles—but no dramatic annual variations. No winter, spring, summer, autumn. No Canada mayflowers or lady's slippers pushing up through winter's leaf litter, transforming the woodland floor. In spring the planet tips its cap sunward. As the wooded path dips toward its exit into the meadow, it passes close to the bank of Queset Brook, where fiddlehead ferns unroll their fronds in sandy soil. Like crosiers they come up, like cudgels, like Irish shillelaghs, shaking their tight little fists at winter past. Then with a confident flare the ferns unroll their broad sails of chlorophyll, drinking up the Sun's red and blue light and leaving the green for the season.

Spring provides a kind of annual recapitulation of the evolution of life on Earth, an opportunity to celebrate anew the greening of the planet three and a half billion years ago by the first photosynthesizing bacteria. All life—the whole glorious parade along the path—depends upon the photosynthesizers. As spring dresses the deciduous woodlands in its Easter best, the nonphotosynthesizers get moving too. Suddenly the woods are skittering, fluttering, munching, singing. From rock-hard seed cases, from underground burrows, from twiggy nests high in the trees, from behind the bark of trees they come, to eat plants, or to eat the creatures that eat plants, or to eat the creatures that eat the creatures. . . . With the invention of photosynthesis, life plugged into a star, and the battle against entropy was won. The universe continues to run down, as it must, but on the surface of the Earth there spreads out a film of highly ordered matter of marvelous complexity and resourcefulness. The one-celled organisms that ruled the

Earth 3 billion years ago were no more advanced than the scum that lives on our shower curtains, but that scum had evolved the ability to make carbohydrates with sunlight. These sunlight-trapping bacteria later lent their talents, as chloroplasts, to the first true plant cells; every cell in every tree in the woodlands contains these organelles that were once bacteria living on their own. Animals developed along a different branch of the evolutionary tree, and it seems unlikely that you and I had photosynthesizers among our ancestors. But the tree of life is a web of interdependence. Green leaves are our necessary link to our yellow star.

ROCKY FOUNDATION

WHEN ARCHITECT Henry Hobson Richardson and landscape designer Frederick Law Olmsted set out to ornament North Easton with native materials, they had no shortage of stone. I stumble over stones as I walk the wooded path, stones humping up out of the soil. Push a spade into the ground at any place in North Easton village, and you will strike rock, of every size and kind. No wonder farmers of the mid–nineteenth century were so willing to abandon this land when better land opened up in the West. No wonder the Ameses were able to assemble so many parcels of tillage and pasturage to build their estates. This rocky soil, called *drift* by geologists, is of glacial origin. Except where the underlying bedrock is exposed as outcrops, almost every particle of stone in North Easton was dropped here by glaciers. For most of the past 3 million years, this part of North America was blanketed by ice, with only occasional warm respites called *interglacials*. At least thirty times great ice sheets accumulated in central Canada and pushed south, eroding the solid crust of the Earth. During interglacials the ice retreated (most recently, 12,000 years ago, not yet to return), melted back toward Canada, dropping whatever eroded material it carried. Every pebble and boulder along my path was abraded or plucked by ice from bedrock north of here, then dropped by melting ice where we find them today. The smaller the stone, the farther north may have been its origin; moving ice slowly grinds its load to dust.

On December 21, 1620, the Pilgrims alighted from the *Mayflower* at Plymouth and according to tradition made their landfall on a rock that has become enshrined in American folklore. Like the Pilgrims themselves, Plymouth Rock is a traveler to the Massachusetts shore, a boulder plucked far to the north by moving ice and dropped at the place where the Pilgrims found it. Erratic boulders—boulders that are dissimilar from the bedrock of the region where they are found—intrigued geologists for a long time before the reality of the ice ages was recognized. Such boulders are common over many parts of the British Isles, northern Europe, and northern parts of North America. In 1825 Peter Dobson of Vernon, Connecticut, proposed that the boulders were carried to their present positions by drifting icebergs at a time when the sea covered the land, presumably during the flood of Noah. According to Dobson, when rock-bearing icebergs drifted south into warmer water and melted, they dropped their passenger boulders in odd places. The drifting-iceberg theory became popular, and erratic boulders, and other alien sediments associated with them, became known as *drift*.

We no longer believe that drift drifted. The largest erratic boulder in New England, and one of the largest in the world, is at Madison, New Hampshire. It is eighty-three feet long and weighs as much as an ocean freighter. A mineralogical examination of the boulder shows that its place of origin was a bedrock outcrop just two miles north of the place where it is presently found. No conceivable combination of rising seas and drifting icebergs can satisfactorily account for the breaking off and transportation of this huge chunk of the Earth's crust. To deepen the mystery, the surfaces of bedrock outcrops in the regions of the boulders are often scratched, and the scratches are invariably aligned in a more or less north-south direction. The erratic boulders always match a source of rock somewhere to the north along the line of scratches.

During the middle of the nineteenth century, an alternative account of the erratics was proposed by Jean de Charpentier, Louis Agassiz, and other astute observers. These scientists guessed that at some time in the past, parts of the northern continents were spanned by vast sheets of moving ice. Exploration of the Greenland Ice Sheet in the 1850s helped convince geologists that continent-spanning glaciers can and do exist. The two-mile-thick Greenland Ice Sheet grows at its center, where it accumulates ice in the form of snow, and moves outward under the influence of its own weight. It is clearly a thing of sufficient power to pick up and carry the ship-sized Madison boulder, and to scratch and polish rocks along its path.

On an outcrop of volcanic bedrock near the path sit half a dozen erratic boulders, some weighing as much as twenty tons, of a coarse-grained pink granite. Once, I chipped off a sample of the rock and followed the bedrock scratches north, looking for the source. Like an Indian trail of bent twigs, the scratches led me several miles out of Easton into the town of Stoughton, where I found what I was looking for, a south-facing ledge of bedrock that under the hand magnifier was identical to the erratics. It is well known that glaciers "pluck" boulders from the downstream side of the rocky outcrops they move across (all of New England's ragged ledges are on the south sides of hills), so I was sure I had found my source. And that's where I stopped. But the trail goes on. The scratches lead north from Stoughton, through the western suburbs of Boston, up the valley of the Merrimack River, veering slightly westward near Concord, New Hampshire, toward the Connecticut River, where they pass into Vermont, and on into Canada.

It is possible to find bits of glacial drift in Easton that had their origin anywhere along the line of scratches. Pieces of New Hampshire, Vermont, and Quebec litter the ground beneath my feet. The trail of scratched rock leads south, too, out of Easton, through Wareham, under Buzzards Bay and Vineyard Sound (which, of course, were not submerged when so much water lay frozen upon the land), to Martha's Vineyard, the southern terminus of the glacier, where even today one might find a scrap of our North Easton bedrock carried there by moving ice.

FOR THOUSANDS of years after the most recent retreat of the ice, New England's glacial deposits were canopied by forests, and a rich soil of organic material built up on top of the clutter of stones. When European colonists arrived and the land was cleared, this fertile topsoil began to wash away. Today, as I walk along the path, I am treading on naked drift, and there is no season when the glacial debris is more obvious than winter. Winter is the time when the stones get up and go, heaving themselves into animation, shaking off the stillness of summer hibernation. They burgeon underfoot like cabbages, shouldering aside frozen earth. From their underground bunkers they push up through frozen soil, threatening to trip the unwary walker. In the meadows sloping down to Queset Brook, they bud from the ground and creep downhill like carapaced horseshoe crabs, unhurried and deliberate.

There is a reason for this winter mobility of stones. It's called *frost heaving* and is caused by alternate cycles of freezing and thawing. The soil surrounding a buried stone freezes and expands, lifting the stone and creating a cavity underneath. Pebbles or grit sift into the cavity. When the ground thaws, the stone is prevented from settling back into its old place. It has been lifted, ever so slightly. Another freeze, another thaw; the cycle is repeated. Millimeter by millimeter the stone makes its way to the surface, finally pushing against tree roots, meadow vegetation, asphalt, or any other obstruction that blocks its ponderous resurrection. Once on the surface, the stone makes its way downhill. Frost lifts a stone in a direction perpendicular to the slope of the hill. Then, as the soil thaws, gravity pulls the stone straight down. Up forward, straight down. Up forward, straight down. Thus do stones descend to the bottom of sloping meadows, taking their sweet geologic time, creeping on frosty fingers.

Robert Thorson, a geologist at the University of Connecticut, has a theory about the ubiquitous stone walls of New England's woodlands. Many of them are waste dumps, he says, for stones heaved up out of the ground by the deep frosts that occurred after the region's trees were cut down in the eighteenth and early nineteenth century. Prior to

the great deforestation, the ground was insulated
from deep frosts by trees and leaves. With the clearing of the trees, stones were waked by repeated
plowing from a subterranean sleep that had been
undisturbed since the end of the Ice Age. As cycles
of freezing and thawing brought stones to the surface, farmers moved them to the sides of their
fields. According to Thorson, many of New England's famous stone walls were not built to fence in
livestock or mark boundaries, but to dispose of
rocks that winter popped out of the ground.

I've taken a close look at the stone walls in our
North Easton woodlands, and there are plenty of
them. The oldest walls, dating from the eighteenth
and early nineteenth centuries, do indeed resemble
heaps of stone. What I had previously considered to
be tumbled-down boundary walls, ravaged by time,
upon closer inspection give every evidence of original chaos, the piles of jumbled stone one might expect if Thorson is right. It has been estimated that
there are more than 100,000 miles of stone walls in
New England, enough to wrap around the world
four times. If only a portion of these are refuse heaps
for unwanted stones, that's still a huge amount of
rock heaved up out of the ground. What I especially
like about Thorson's theory is the life it gives to
stones, lifted from Ice Age graves into sunlight to become the bane of eighteenth- and nineteenth-century farmers and—piled into walls—a charming

signature of New England and habitat of chipmunks and lichens.

BEFORE THE WOODED PATH debouches into the meadow, an opening in the trees provides a brief vista out across the valley of the Queset Brook to the place where the Sheep Pasture mansion once stood. The house is gone now (cheaper to demolish than pay taxes), but the elevated terrace still stands like a rocky fortress. If I leave the path, cross the brook, and walk along the terrace's retaining wall, it is like being next to a natural cliff. The wall is constructed of roughly dressed stones, some quarried from nearby outcrops, others—glacial drift—presumably dug out of the ground as the builders excavated the mansion's foundations. Along with the local pink granite and green country rock are kinds of stone not native to North Easton. The sources of these alien stones are north of here, in the town of Stoughton or even farther afield, somewhere back along the line of scratches.

I wonder if the four children—Elise, Olivia, Oliver, and Richard Ames—who grew up in the Sheep Pasture mansion early in the last century were taught the lessons of the ice ages. Certainly, the grown-up memoir of one of those children, Elise, tells of pony-cart rides with the family coachman, John Swift, who taught the children much about the natural history of the estate. Did he tell them the story of the traveling stones? The children surely played on the towering crag of exposed bedrock just across the driveway from the mansion's entrance. Perched atop the crag is a car-sized boulder of Stoughton granite, which, when my own children played there in the 1960s, was known as Dogface Rock, for that is what it looks like when viewed from a certain angle. I climbed up there recently; the "dog face" is still recognizable, as it must have been to the Ames children a hundred years ago, and as it will be to other children generations hence. It endures where it was dropped by melting ice, with geologic repose. As I stand next to it, tracing the features of the "dog's" mouth, chin, nostril, and eye, I am swept back thousands of years to that former world of continent-spanning glaciers. It was to the detritus of that deep geologic history that Olmsted applied the shaping force of human imagination, here at Sheep Pasture and throughout North Easton, to create artificial landscapes that pay homage to the deep history of the land.

Olmsted's plans for the Sheep Pasture estate, collected at the Olmsted Archives in Brookline, Massachusetts, begin with a preliminary survey of the land as he found it: natural boundaries, existing stone walls, trees worth saving, outcrops of bedrock. This "lay of the land" was the canvas upon which he would work. In his earliest sketches, one can sense him playing with the landscape, imagining where the carriage roads might be, what might be the visual prospect from each turning in the roads, how the woodlands and the watercourses might be shaped, where a hill might be trimmed or augmented. Of course, most important, was the siting of the house. On high ground, certainly, with fine outlooks from the terrace. One drawing is a cross-section of the land falling from terrace to brook, a sight line out over the parapet's capstones of Cape Ann granite; the architect clearly had in his mind's eye exactly what would be the view from each corner of the terrace, and even from the doors and windows of the house, long open vistas framed by water and trees. With his fellow landscape architects of the late nineteenth century, Olmsted sought to create something that was not quite wilderness and not quite civilization, but a hybrid of art and nature. He had no qualms about moving tons of earth, diverting streams, clear-cutting trees, or planting forests if the result was beautiful and ostensibly "natural." One hundred years later it is impossible to distinguish at Sheep Pasture the works of man and nature. I stand on the terrace of the vanished mansion and look out upon a landscape that expresses the ideal of the early-twentieth-century landscape theorist Henry Vincent Hubbard: a landscape "plainly once the work of man, but so far received back by nature that man's interference is no longer an incongruity, but rather an added pleasure of association."

Olmsted and his colleague Charles Eliot, designer

of Boston's metropolitan system of parks and park-ways, thought of themselves as educators as well as architects. (Eliot was the son of Harvard University's president; it was perhaps inevitable that he would perceive a pedagogical element to his art.) Their audience was the harried urban middle class, their message "the beautiful and reposeful sights and sounds which nature, aided by the landscape art, can abundantly provide." In this, they were almost certainly right. Try to imagine New York City without its Central Park, or Boston without its Middlesex Fells and Blue Hills Reservations; the prospects are unthinkable. Turn-of-the-century landscape designers hoped to convey such qualities as "Life, Power, Beauty, Peace, Joy, Mystery and the Holy Spirit," to quote landscape gardener Frank Waugh, which sounds a bit overblown to us, but there is certainly an element of truth to their perception of the ameliorating influence of nature. Not wild nature, of course, but nature tamed by conscious intent. An artfully contrived landscape reveals nature's essential lesson as well as any text of Emerson or Thoreau, they might have said.

And what is nature's lesson, as understood by the educated offspring of the nineteenth-century entrepreneurial class? That we are part of an organic world, and that we need, as Olmsted insisted, "relief from the too insistently man-made surroundings of civilized life." When Ames coachman John Swift took his young charges, Elise, Olivia, Oliver, and

Richard, along those sections of carriage road that are now part of my path, he was doing more than teaching them the names of plants and animals; he was giving them a context in which to understand that human life, although rooted in a particular place and time, is not bounded by the here and now, and that spiritual wealth comes in currencies that cannot be accumulated in Boston banks.

FROM THE CLEARING in the woods that gives a glimpse of the terrace of the now-demolished mansion, the path reenters the trees for another few hundred paces, then emerges onto one of the estate's carriage roads not far from where the road crosses the brook. Here the lay of the land changes dramatically. The rolling uplands that endow the Queset Brook with its gravitational force give way to low-lying plains and water meadows. The ground underfoot changes as well. No more boulders bulge up out of the ground to trip the unwary walker. Now the soil becomes sandy and fine, and a spade can be driven deep without the resounding New England *clunk*. A glance at a geology map reveals why. North of a line that runs east-west along the carriage road, the underlying bedrock is igneous: granite and a tough "country rock" of volcanic origin, both resistant to erosion. South of the carriage road, the bedrock is more easily eroded sandstone. During the millions of years that the glaciers moved across the land, they abraded the sandstone more deeply than the granite and country rock. Then, at the end of the Ice Age, retreating ice dropped its burden—boulders, pebbles, powdered stone—everywhere, on the resistant igneous uplands and on the more deeply eroded sandstone lowlands. Immediately, meltwater flowing away from the edge of the shrinking glacier carried away the pulverized rock from the uplands, leaving behind the characteristic drift of larger stones, and deposited the silty detritus in the lowlands. There are just as many large stones underfoot in the sedimentary lowlands, but they are buried deep under a blanket of sand and fine gravel washed down from the highlands. Geologists call this blanket of fine glacial debris an *outwash plain*.

And so the Sheep Pasture acres are of essentially two kinds: igneous and sedimentary, high and low, stony and fine. The house, naturally, would be situated on the high land, with commanding prospects out across the valley of the Queset Brook and the outwash plain to the south. The same elevated siting was applied to other third- and fourth-generation Ames mansions in North Easton, and to the mansions of the wealthy sons and daughters of entrepreneurial industrialists in other suburbs of Boston, New York, Philadelphia, Pittsburgh, and Cleveland. There is a symbolism in the siting, of course—it spoke of bedrock bourgeois values: practical, sober, plain-speaking, democratic, financially secure. To build on rocky crags suggested an irrepressible confidence in the power of money and technology to reshape any obstacle of nature, the same confidence that enabled Oliver II and Oakes Ames to throw a railroad across the continent (despite more than a whiff of scandal).

My place of work, Stonehill College, was another Ames estate astride the boundary between igneous uplands and sandstone lowlands. When the founders of the college bought the estate in the 1940s, they installed themselves in the grand mansion on the hill, called Stonehouse Hill House, with a foundation blasted from bedrock. As the college expanded, new construction moved downhill onto the outwash plain. To build on flat sand and gravel was cheaper; less aesthetically pleasing, perhaps, but certainly easier to fit the spartan buildings with their foundations and septic systems into the easily excavated and permeable lowlands. After all, this was a poor college, catering to the sons (and later daughters) of the working class. Today, little of Olmsted's sort of nature romanticism can be seen on the college's lower campus, merely workaday utility. The "Big House" on the hill continues to dominate the campus, physically and aesthetically. Although the college prides itself on serving the children of the less affluent classes, the glossy photographs on the covers of its catalogs and brochures invariably feature the colonnaded upland mansion as the proud symbol of the college, not the more prosaic buildings on the outwash plain. Proletarian aspirations have co-opted blue-blooded grandeur.

Olmsted and his contemporaries built estates for the wealthy, but they created great public spaces too. When Charles Eliot decided to ring Boston with a system of metropolitan parks, he caused to be immediately acquired the prominent uplands of the Middlesex Fells north of the city and the Blue Hills to the south, lofty tracts of land that might otherwise have become residential precincts of the wealthy, but which are now ornaments serving all the people. Many private estates created for wealthy clients by Olmsted and his disciples are now in public or semi-private possession, as at Sheep Pasture and Stonehill College, so that every day my path takes me through terrain that reflects American bourgeois values in service to the common weal.

If public spaces of the quality of Olmsted's Emerald Necklace of Boston parks and Eliot's Middlesex Fells and Blue Hills Reservations are no longer being created, and if (as seems to be the case) those precious public spaces are being nibbled away by superhighways and neglect, it is perhaps because we no longer muster the mix of public spirit and aesthetic purpose that marked that earlier generation of landscape designers. For all of their Brahmin pride, these men dedicated themselves to a vision that transcended their privileged origins. In an 1891 statement to Boston's Trustees of Reservations, Eliot said: "Here is a community, said to be the richest and most enlightened in America, which yet allows its finest scenes of natural beauty to be destroyed one by one, regardless of the fact that the great city of the future which is to fill this land would certainly prize every such scene exceedingly, and would gladly help to pay the cost of preserving them today." His contemporaries responded to his plea with creative determination, and a great system of public parks and reservations was the result.

THE GLACIAL DRIFT beneath my feet along the path, those chunks and particles of northern Massachusetts, New Hampshire, Vermont, and Quebec—those nocturnally active boulders that humped down frosty slopes and found their way into the characteristic stone walls of New England and the terrace of the Sheep Pasture mansion—may owe their presence here to colliding continents half a world away.

Millions of years ago India, drifting northward on the mobile surface of the Earth, nudged into Asia and began pushing upward a double-thickness slab of Earth's crust known as the Tibetan Plateau, the front range of which are the towering Himalayas. Sunlight baking down on the high plateau causes the air to warm and rise. To take its place, moist air sweeps in off the Indian Ocean, part of a great convection cycle known as the Indian monsoon. These rag-wet winds climb the southern flank of the Himalayas and release their burden as drenching rains. The water combines chemically with the rock, weathering it, cutting the mountains down. The chemistry of weathering takes carbon dioxide out of the air, substantially reducing the amount of that gas in Earth's atmosphere, not just in Southeast Asia but, because the atmosphere is well mixed by winds, everywhere. Carbon dioxide is a *greenhouse gas*; it traps the Sun's heat at the surface of the Earth. As the mountains rose and eroded, the amount of carbon dioxide in the atmosphere decreased and Earth began to cool. On northern continents ice sheets formed and moved across the land, grinding, abrading.

Astonishingly, *according to this theory,* mountains in Asia caused glaciers in New England. I tell this story for two reasons: First, it suggests that every local environment must be considered in a global context; and, second, my daughter, Maureen E. Raymo, a geologist at Boston University, contributed significantly to inventing the theory and winning it widespread acceptance.

The glacial chill was not constant. Periodic wobbles in the Earth's orbit vary the budget of solar energy falling upon the Earth and caused the ice to periodically recede, roughly every 100,000 years; this accounts for the so-called interglacials. We are in an "interglacial" now, except this time the ice may not return as it did dozens of times before with relentless regularity. This time a new creative force is at work on the planet. Human ingenuity—buttressed by scientific knowledge and technological prowess—is stronger than the geologic and astronomical forces that have worked on the face of the globe from time immemorial; even the climate is subject to our whim.

LANDSCAPE DESIGNERS at the turn of the last century sought to *engineer* landscapes that manifest "Life, Power, Beauty, Peace, Joy, Mystery and the Holy Spirit," those qualities of refined nature listed by the early-twentieth-century landscape gardener Frank Waugh—a tall order, surely, but to that end they cut and dredged and shaped and filled and planted. So much of what they sensed as valuable in a landscape was "rapidly vanishing for all eternity before the increased thoroughness of the economic use of land," wrote Frederick Law Olmsted's son and successor, Frederick Law Olmsted Jr. That the Olmsteds and their contemporaries successfully resisted the economic tide is evidenced by the public and private spaces that remain today, a century later, the greatest adornments of the continent.

I stand by the terrace wall of the now-vanished Sheep Pasture mansion and wonder what this massive structure has to do with the qualities Waugh wanted to manifest in nature. Well, *power* maybe, and certainly great family wealth, but how does the imperishable materiality of stone evince spiritual values? What is the metaphorical lesson of geology here articulated by the landscape's art? Back in the

1940s and 1950s when I was in school—parochial school, staffed by nuns of Irish immigrant stock—materialism was portrayed as the mighty foe of every spiritual value. In my youthful innocence, I was never quite clear what materialism was. My teachers might as well have said "Beelzebub," and indeed they presumably meant something of the sort. Whatever materialism was, it clearly made no place for spirit. Whereas the Protestant ascendancy emphasized the spiritual rewards that would inevitably flow from mastery of the material world, my parochial school teachers reversed the equation: Cultivate the spirit here on Earth, they said, and you'll find your material reward in Paradise. Let the "materialistic" Protestants have their colonnaded mansions on granite crags; we spiritual Catholics would wash our souls in the iffy promises of the outwash plain.

These opposing philosophies of matter and spirit presumably came from our ancestors' experience of the world. Materialism may have had its origin in the experience of thunder and lightning, Sun and Moon, weight and force, the sharp edge of a knapped stone, fire, food, blood, bone. The idea of spirit may have sprung from self-awareness, dreams, light and dark, the mysteries of birth and death, the vague but incontrovertible intuition that there is more to the world than meets the eye.

Olmsted and Eliot had considered the same categories of matter and spirit fifty years before my teachers presented them as irreconcilable opposites. In designing landscapes such as Sheep Pasture, Olmsted and Eliot sought to bring the two together: material nature made to manifest transcendent realities. And that, I think, is what the Sheep Pasture mansion's massively walled terrace is all about. As Olmsted's functionaries rolled the great boulders into place to make a satisfactorily imposing pedestal for his client's home, they undoubtedly imagined that in their art-inspired labors machinery served mystery, and vice versa.

Early in the twenty-first century, the old dualism of matter and spirit seems irrelevant. In the new theories of physicists, the fundamental material particles—protons, neutrons, electrons—dissolve into a kind of cosmic music, all resonances, vibrations, and spooky entanglements. Matter has revealed itself as a thing of astonishing, almost *immaterial* subtlety. The one property of matter that lingers is its *potentiality*. The hydrogen and helium atoms forged in the Big Bang possessed a built-in capacity to complexify and diversify, to spin out stars and galaxies, carbon, silicon, oxygen, iron, and ultimately the rocky substrate of the Earth, even life and consciousness. Far from explaining away the mystery of the world, our new knowledge of matter rubs our noses in mystery. The more we learn, the more we become aware that matter—ordinary matter, even the handful of glacial drift that I scoop up along the path—is more than we had ever dared to guess.

The great landscape architects of the turn of the last century wanted *things* to show forth the ideal. As I walk a path haunted by Olmsted's spirit, I am cognizant that the dust under my feet contains sufficient mystery to occupy me for another lifetime of study, spiritually resilient in its unabashed materiality—art and nature working hand in hand.

VERGES

IN SHEEP PASTURE'S heyday, Oliver Ames employed eight men to maintain the grounds. Olmsted's original horticultural plan for the estate suggests something of the task that confronted the gardeners: barberry, hydrangea, spicebush, spirea, periwinkle, sweet pepper bush, swamp pink, mountain laurel, ash berry, bittersweet, quince, *Rosa multiflora*, silver bell tree, and dozens more species are listed. (Periwinkles have colonized one wooded area along the path; I wonder if they are escapees from the original gardens.) It's hard to know how many species of the original plan were actually planted, for wild species again have almost exclusive possession of the property.

As I leave the woods and step onto the carriage road that borders the meadow, I pause to observe one of spring's most appealing wildflowers, the wood anemones, tentatively unfolding their five-petaled blossoms as if testing the temper of the air. They are small, inconspicuously colored, and inclined to bashfulness, hesitating in the woody verges like young ballerinas waiting in the wings for a more colorful prima donna to take the stage. Later, there will be more conspicuous floral displays—open fields of raw gold, purple marshes, towering spikes of crimson—but no wildflowers are more welcome than the unassuming spring pioneers of the shady verges. Wood anemone, bellflowers, starflowers, false Solomon's seal; these ornament the path as it emerges from the woods.

The Ames children, Elise, Olivia, Oliver, and Richard, learned the names of Sheep Pasture's wild-flowers from their nurse, Matilda Golden, who painted each species in an album and labeled it. She might have used as her guide Mrs. William Starr Dana's *How to Know the Wild Flowers*, published just before the turn of the twentieth century by Charles Scribner's Sons. Subsequent field guides are more comprehensive, authoritative, and up-to-date, but none captures the delicacy and charm of Mrs. Dana's original. She was born Frances Theodora Smith in 1861 and brought up in New York City. Her love for wildflowers was acquired during summers spent with her grandmother in Newburgh, New York, not far from the home of the famous naturalist-writer John Burroughs. While in her early twenties, Frances Smith married William Starr Dana, a naval officer much older than herself. The marriage was happy, but the husband soon died of influenza. Victorian convention dictated a long period of mourning, widow's weeds, and retirement from society. As a distraction from grief and social inactivity, Mrs. Dana took up again her childhood interest in natural history.

In those days, the only sources of information about wildflowers were technical works such as Asa Gray's *Manual of Botany*. Field guides for the casual observer simply did not exist, and if one had a nanny with some traditional knowledge of flora, as did the Ames children, one was lucky. But the need was there, and Mrs. Dana found her inspiration in a magazine article by her old neighbor John Burroughs. "Some of these days," wrote Burroughs, "someone will give us a handbook of our flowers, by the aid of which we shall all be able to name those we gather on our walks without the trouble of analyzing them. In this book we shall have a list of all our flowers arranged according to color, as white flowers, blue flowers, yellow flowers, red flowers, etc., with the place of growth and the time of blooming." Mrs. Dana took up the challenge and created a handbook that was a model for all that came after. Her friend Marion Satterlee supplied delicate pen-and-ink drawings to complement the text. The first printing sold out in five days, and the clothbound book stayed in print until the 1940s.

How to Know the Wild Flowers has recently been reissued in a collector's edition, keeping it in print for a century, surely a record for a guide of this sort.

What gives Mrs. Dana's book its enduring charm are the brief essays describing each flower, written in the best tradition of Victorian natural history—warm, literate, anecdotal—the ideal botanical accompaniment to Olmsted's landscape architecture and the floral and faunal motifs with which Richardson decorated his buildings. Mrs. Dana frequently quotes Greek and Roman authors, Shakespeare, Wordsworth, and of course the New England poets and essayists—Longfellow, Whittier, Thoreau, and Emerson; her inspirations were the same as those of the architects who ornamented North Easton and designed the Ames estates. To read her book is not only to learn the wildflowers but also to stroll down the primrose path of cultural history to a time when wild plants and animals were more a part of everyday experience than they are today.

What did Elise, Olivia, Oliver, and Richard know of the wood anemone? The name means "wind-shaken," according to Mrs. Dana, who provides a snippet of William Cullen Bryant: "Within the woods, whose young and half transparent leaves scarce cast a shade, gay circles of anemones danced on their stalks." A dollop of Whittier, too. And then some flower lore from ancient Greece that would have changed forever how the children perceived the wood anemone where they found it blooming in the sun-dappled shade: In Greek lore the flower sprang from tears shed by Venus over the body of her slain lover Adonis.

But Mrs. Dana's handbook is not all quaint Victorian charm. She also tells us *when* to expect the flowers and *where* to find them. She tells us their Latin names and botanical families, and describes in brief nontechnical words the form of stem, leaves, and blossoms. All of this can be found in modern guidebooks, too, but what the newer books lack is Mrs. Dana's personal, literary touch that makes *How to Know the Wild Flowers* worth owning a century after it was written. Natural histories such as hers remind us that it is possible to find, with Shakespeare, "tongues in trees, books in running brooks, sermons in stones, and good in everything."

THE WOOD ANEMONES that hunker in the shadows are native to North America. But many other wild plants along the path are newcomers to the continent. Purple loosestrife, Queen Anne's lace, lady's thumb, butter-and-eggs, catchfly, hawkweed, chickory, and the common dandelion are just a few of the wildflowers that made their way to our shores by crossing the Atlantic. Some of these plants probably arrived on the *Mayflower*. No sooner had Miles Standish and his friends stepped down from Plymouth Rock than chickweeds, mulleins, and stinging nettles sprang up at his feet. (It was Miles Standish who deeded to an Ames ancestor the plot of land in West Bridgewater, Massachusetts, where the family first manufactured shovels.) Native Americans called plantain "Englishmen's foot" because it seemed to magically appear wherever the colonists walked. As early as 1672, the natural historian John Josselyn recorded forty species of weeds that had "sprung up since the English planted and kept cattle in New England." Daisies, bouncing bet, and most buttercups came with the cows, and although some might call them weeds, they add a lovely old-world charm to our fields and verges. The majority of plants in my Peterson's wildflower guide are immigrants. Out of the holds of ships, on the hides of cattle, in the seeds of crop plants, in bedding straw, in foodstuffs, snagged in the cloth of Miles Standish's cloak and stockings, they came, the seeds of English plants, to find new worlds to conquer.

Many plants new to North America first sprouted up alongside wharves and shipyards. From there they made their way inland along new roads hacked out of the wilderness, and later along canals and railroad embankments, taking up residence in any sort of disturbed soil. Native plants adapted to quiet precolonial forests and meadows gave little competition to the aggressive intruders. As Pilgrims and Puritans leveled the ancient New England forests, their floral co-colonists thrived in a landscape that soon came to resemble that of their former European home. Plants that thrive in verges, especially, came into their own.

Traffic in the opposite direction was not nearly as successful. Seeds that made it eastward across the Atlantic as stowaways in vessels were met at European shores by native plants long adapted to their cultivated environment. The American newcomers didn't have a chance of competing. But European horticulturalists had a keen interest in plants of the New World and made places for them in gardens and botanical collections. The earliest explorers of the North American continent included avid natural historians who collected plants and animals for shipment to Europe, an often ill-fated venture. Consider the misfortune-prone Scot, John Goldie, who vigorously botanized in Canada and the northern United States at about the time the Ameses began their shovel-making operations in North Easton. His efforts to inform the Old World of the treasures of the New were fraught with uncertainty. He put his hard-won specimens on sailing ships bound for home; one, two, three collections went astray, one in a shipwreck, two vanishing in transit. Sea captains responsible for cargoes of economic importance could hardly be expected to pay much mind to packages of roots and seeds. Ever persistent, Goldie finally managed to get a collection of plants back to Scotland.

In this saga of two-way transatlantic traffic of flora and fauna, there is an interesting story involving

Thomas Jefferson that demands telling. It began with Georges-Louis Leclerc de Buffon, the world's most respected zoologist and botanist during the reigns of the French monarchs Louis XV and Louis XVI. Buffon was something of a sycophant, a sometimes slipshod scientist, and a protégé of Louis XV's supremely influential mistress, Madame de Pompadour. At the wishes of his royal masters, Buffon stocked the park at Versailles with wild animals. He restored and expanded the Jardin des Plantes on the Left Bank in Paris. And with the confidence that goes with being a favorite of kings, he embarked upon writing a comprehensive natural history of the world. In that work, Buffon expressed the pontifical opinion that native animals of the New World are smaller than those of Europe, that there are fewer species of animals in America, and that even domesticated species of European animals (the cattle, sheep, and swine of the colonists) diminished in size and vigor upon crossing the Atlantic. The native peoples of America shared this diminutive tendency, wrote Buffon, possessing "small organs of generation" and "little sexual capacity."

On his mountaintop in Virginia, Thomas Jefferson read these opinions with outrage. In his *Notes on the State of Virginia,* he went to great lengths to refute his scientific rival. He compiled weights of North American and European species in an attempt to show that new-world animals are bigger, more varied, best. None of this had any effect upon Buffon, but it undoubtedly made Jefferson feel better. Jefferson was not alone in being offended by Buffon's blatantly chauvinist pronouncements. The Frenchman's name was a favorite target for denunciations at Fourth of July celebrations throughout the young American republic.

In 1784 Jefferson was appointed minister to France. Before leaving for the continent, he traveled to Boston, where he purchased an enormous panther skin with which to regale the eminent French naturalist. When at last the two men met at Buffon's home in the Jardin des Plantes, Jefferson offered his friendship and his statistics regarding the relative vigor of American animals. Buffon was unfazed. He took down a copy of his latest work from the shelf and said to Jefferson, "When Mr. Jefferson shall have read this, he will be perfectly satisfied that I am right." Jefferson thereupon produced his panther skin, momentarily confounding his adversary.

Now Jefferson waited to deliver his coup de grace. While still in America, he had asked John Sullivan, governor of New Hampshire, to capture a moose, the biggest that could be found, prepare it for stuffing, and ship it to Paris. The governor's agent thereupon "sallied forth with his forces" into the snowy wilderness of Vermont, where he killed "with Difficulty" a moose. It took two weeks to remove the animal from the forest, a task that required building a twenty-mile road to the nearest settlement. By the time the moose reached Governor Sullivan in the state capital, it was already in a state of putrefaction. Sullivan set about having the moose cleaned and prepared for shipment, a job (as he wrote Jefferson) "such as was never before attempted." The moose's antlers were apparently unimpressive, but Sullivan sent along the horns of a deer, an elk, and a caribou. "They are not the horns of this moose," he wrote, with some lack of scientific objectivity, "but they may be fixed on at pleasure." He understood the impression Jefferson was intent on making.

Sullivan's bill for services rendered arrived in Paris before the moose. Jefferson had expected that an animal could be obtained from some hunter for a pound or two, to which might be added a few pounds for shipment. To his astonishment and dismay, Sullivan's carefully itemized invoice amounted to forty-six pounds sterling. As nearly as I can figure it, this would be about $4,000 to $5,000 in present money, a pricey putdown for Buffon. At last the moose itself appeared in Paris, in an appalling state of decay. A good part of the hair had fallen out in transit, and the carcass probably smelled to high heaven. Jefferson nevertheless sent it on to Buffon, along with the horns of the elk, deer, and caribou, assuring him that all of the specimens were disappointingly small. "The horns of the deer which accompany these spoils [sic], are not the fifth or sixth part of the weight of some I have seen," wrote Jefferson in his cover letter, undoubtedly with some red-faced embarrassment. Nevertheless, he toughed it out. Buffon was gracious but unimpressed. His book went unrevised.

Transatlantic competition did not cease with Jefferson's contretemps with Buffon. As the Sheep Pasture estate was being built at the end of the nineteenth century, the English naturalist Alfred Wallace commented in the *Fortnightly Review,* with Buffonesque confidence, that nowhere could be seen in America such brilliant masses of wildflowers as are displayed in Britain's moors and meadows. Mrs. William Starr Dana rose to rebut Wallace in her second book, *According to Season.* In autumn, she wrote, she'd put New England's asters, goldenrods, and joe-pye-weeds up against anything the British countryside might offer. By Mrs. Dana's time, it was the Americans who were collecting abroad, looking for exotic species to bring into their landscaped gardens. Olmsted's horticultural plan for Sheep Pasture lists among species to be planted Indian currant, Persian lilac, California privet, Japanese globeflower, Tartarian honeysuckle, English ivy, Chinese wisteria, and Labrador tea, suggesting a certain catholicity of taste. Charles Sprague Sargent, friend of the Ameses and director of Boston's Arnold Arboretum, sent his agents to every temperate continent to gather trees. (My favorite at the arboretum is the dove tree from China, which in spring seems to be covered with fluttering white hankies.) And, of course, he hired the indefatigable

Olmsted to lay out the grounds of the arboretum.

Olmsted had been inspired to a career as a landscape architect by visits to England as a young man,
particularly to the royal gardens of London and
Birkenhead Park near Liverpool. At the Ames estates in North Easton, as at Arnold Arboretum and
other public parks and private estates, he sought to
recreate in America what can only be described as
English countryside. Many of the wildflowers along
my path, immigrants from Britain, are perfectly
adapted to an environment that now resembles
their former home. By the time Olmsted sculpted
Sheep Pasture into an attractive ensemble of rolling
woodlands and meadows, our native moose had
pretty much gone the way of bears and panthers, so
even the animal population was not dissimilar from
that of Europe. It may be that in some Heaven reserved for naturalists Buffon had a last-last laugh at
Jefferson's expense. Whatever the relative virtues of
native European and American fauna and flora, it is
more or less a European countryside that I walk
along my path today—a far cry from the thickly
forested landscape of precolonial America.

FROM WHERE THE PATH exits the woods, it is
only a few long strides along the carriage road to the
plank bridge over Queset Brook, and here in the autumn are gorgeous displays of purple loosestrife, a
plant that Mrs. Dana called a "radiant acquisition"

from foreign shores. We can forgive the old country for sending us so many weedy pests, she said, when one has seen a marsh aglow with this beautiful plant. As she rightly inferred, one person's pest is another person's miracle, and sometimes vice versa. Weed or adornment, the loosestrife's remarkable blossoms will reward anyone's close observation with a magnifier.

The plant's sexual organs, male stamens and female pistil, come in three arrangements: short and midlength stamens, and long pistil; short and long stamens and a midlength pistil; or midlength and long stamens and a short pistil. All blossoms on any one loosestrife plant are of the same arrangement. Now here's the clever thing, the thing one wouldn't know without a bit of careful study on the part of botanists or horticulturalists: A plant can only be fertilized when pollen from a male organ lands on a female organ *of the same length*. For example, a plant with a midlength pistil can only be fertilized by a plant with midlength stamens. And so on. Thus no blossom can be fertilized by another blossom on the same plant. This guarantees cross-pollination between plants, which confers evolutionary advantages. (Cross-pollinated plants are often better adapted to survival and reproduction than either parent, and they avoid the genetic deterioration that sometimes results from inbreeding.) I marvel at the scientific work that must have been necessary to discover and confirm the purple loosestrife's curious reproductive strategy. There still remains the untold story of the lock-and-key molecular chemistry that regulates loosestrife fertilization, and of how the chemistry is controlled by genes. On the molecular scale, the selective fit between sperm and egg must be a thing of almost unbelievable subtlety and refinement.

The big question, of course, is how such cleverness in nature comes about. That such refinement of design could have resulted from random mutations, as described by Charles Darwin, seems—well—incredulous. We have heard the argument again and again: A thousand monkeys banging on a thousand typewriters for a thousand years could not produce the works of Shakespeare. They could not even produce those three wonderful sentences: "But soft! What light through yonder window breaks? It is the east, and Juliet is the sun." Botanically speaking, say the critics of Darwinism, random mutations are the thousand monkeys banging on typewriters and the loosestrife's sexual subtlety is *Romeo and Juliet*.

The fallacy here is the assumption that evolution is based solely on random mutations of genes, which is not the case. The thousand-monkey metaphor is an inadequate—indeed, perverse—misrepresentation of Darwin's great idea. Yes, there is a random or capricious element in the theory of evolution as it is currently applied. Genes mutate, either during the hugely complex process of reproducing themselves or by environmental transmutation. There is no denying this; such changes have been exhaustively demonstrated. And for all practical purposes, the changes *appear* haphazard. But Darwinian evolution is not a willy-nilly accumulation of random mutations. Genetic changes are selected for their survival value in a particular environment. This, too, cannot be doubted; nature is manifestly relentless in its harvest of the weak. The evolution of organisms by natural selection is not just a theoretical possibility; it is inevitable. It would require the intervention of a divine being to keep it from happening.

Back to the monkeys. Let's say that every time a monkey types by chance a string of *A*s I give it a banana. And I give the other monkeys a boot in the behind. Pretty soon we'd have a lot of monkeys hitting the *A* key incessantly. And no one would call it random. Selection has been at work. If I had supervision of a thousand monkeys not for a thousand years but, say, for a million generations, a sufficient supply of bananas, and a durable boot (the well-fed, unbooted monkeys being most successful at procreation), sooner or later I would almost certainly witness a creature who sat at his or her machine and typed: "Arise, fair sun, and kill the envious moon, who is already sick and pale with grief that thou her maid art far more fair than she." And the same dynamic applies exactly to the evolution of organisms, where competition for scarce resources provides the bananas or boot.

Computer simulations of evolution have shown

that even such complex apparatuses as the loose-strife's ingenious sexual organs can arise with surprising celerity with an appropriate mix of random change and ruthless selection. The universe has built within it an endlessly creative capacity for self-organization, and every plant, every creature along the path has a story to tell of increasing diversity and complexity—tongues, books, sermons, and good news in everything. As I get down on my knees and examine the loosestrife's astonishing sexual apparatus under the magnifier, I am attending with rapt amazement to the fundamental story of creation: evolution by common descent and natural selection. John Burroughs, who inspired Mrs. Dana's guidebook to the wildflowers, wrote: "Most young people find botany a dull study. So it is, as taught from the text-books in the schools; but study it yourself in the fields and woods, and you will find it a source of perennial delight."

WHERE ALONG the trajectory of cosmic evolution lies the strange sexual apparatus of the purple loosestrife? Where lies human curiosity? What *is* the most complex thing in the universe? In a universe of hundreds of billions of galaxies (at least), does the level of complexity we encounter here on Earth stand high, low, or somewhere in between? And does it matter?

The physicist Eric Chaisson tries to answer these

questions in his book *Cosmic Evolution: The Rise of Complexity in Nature*. He develops the idea of "time's arrow," a curve of rising cosmic complexity that begins with the Big Bang and reaches—well, insofar as we know and for the time being—a brain that can appreciate the subtlety of the loosestrife's sexual equipment or the beauty of a wood anemone in a shady verge. He argues that rising complexity can be explained (or at least roughly described) by the laws of *nonequilibrium thermodynamics*, a well-established branch of physics, without any need to postulate the action of an ever-intruding deity. He shows that in an expanding universe, local pockets of order will inevitably arise even as the overall disorder (entropy) of the universe increases. There is nothing new in any of this; scientists have long known that a growth of local complexity is entirely consistent with the laws of thermodynamics. What is most original about Chaisson's argument is his proposal of a way to measure complexity, and to plot the course of cosmic evolution.

Chaisson invokes something called *free energy rate density*—defined as the amount of useful energy contained within any quantity of mass and interval of time. For example, vastly *more* energy flows through a star than is involved in a worm's metabolism, but the *concentration* (density) of energy is greater for the worm than for the star (roughly 10,000 ergs per second per gram for a worm, versus 2 ergs per second per gram for the star), and therefore, according to Chaisson, the worm is more complex than the star. As the universe evolves, the complexity of things increases: stars, planets, plants, animals, brains, societies. Within each of these broad categories, the free energy rate density climbs, as, for example, within human culture from hunter-gathering, to controlled use of fire, to agriculture, to industrialization. Chaisson takes this rising curve to be the signature of cosmic evolution. We have seen this successive augmentation of free energy rate density along the path as Native Americans, European colonial farmers, and entrepreneurial industrialists extracted ever more energy resources from these few square miles of Earth's surface in ever briefer intervals of time.

This certainly does not imply that nineteenth-century industrialists led better, more fulfilling lives than smallholding farmers, or that colonial farmers led better, more fulfilling lives than Native Americans. But it does suggest that the cosmos naturally breeds complexity. It also suggests that there is an inevitability to time's arrow that is fruitless to resist. No matter how much we might idealize the Native American's relationship with the land, or the charming simplicities of Thoreau's farmsteaded countryside, there is no going back to a hunter-gatherer way of life or to the homespun economies of Concord, at least not for the vast majority of the 6 billion of us who currently inhabit the Earth. And an even greater concentration of free energy will be necessary if we are to feed ourselves in our teeming billions, keep ourselves healthy, and provide a modicum of creature comforts.

Where will evolution take us from here? If we accept Chaisson's analysis, at least tentatively, then certain conclusions follow. First, we are almost certainly not unique in the universe, because there is nothing in his account of cosmic evolution that evokes the uniqueness of the Earth. And second, we are not the end of the line on Earth; the curve of cosmic complexity will continue to rise, and further novelty will ensue. A glimmer of the future might already be visible. Presently, in our neighborhood of the cosmos, the top of Chaisson's curve of evolution is occupied by—get this—the computer microchip, with a free energy rate density of something over 10 billion ergs per second per gram. This exceeds even the human brain because microchips perform their operations so much faster than do webs of organic neurons. Computer chips are not as sentient as human brains, but they are more complex by Chaisson's definition, and represent a further step in cosmic evolution by his standard. Does the immediate future of complexity on this planet lie in dense webs of silicon transistors? Are humans merely instruments in bringing about the next step in cosmic evolution? Are there other forms of complexity out there among the galaxies with free energy rate densities in the hundreds of billions? Time will tell, although I would not be surprised if the answer to all of these questions is yes.

This much is certain: The future of the planet will not be a reprise of the past, a return to "a state of nature." The future will certainly be technological, increasingly globally homogeneous, and, in the short run at least, will embody the connectivity of the computer chip and the contrivances of genetic engineering—in conformity with Chaisson's law of rising complexity. American conservationists frequently offer Native American attitudes toward nature as the solution to our environmental dilemmas, and certainly there is much we can learn from Native American wisdom, as we can learn from the wisest voices of every tradition. But the precolonial North American "wilderness" was not so pristine as conservationists sometimes imply, nor were the original human inhabitants of the continent so peaceable in their relationships with the native fauna and with each other. In any case, there's no going back. Whatever unspoiled paradise European colonialists thought they found in the New World has long been erased by human cunning. Within 400 years, the environs of the path have undergone transformations from hunting-gathering to agricultural to industrial to cybernetic, following the curve of rising complexity. Many North Easton villagers now make their living in front of a computer screen. They are by and large more prosperous, healthy, and leisured than at any time in the past.

I am constantly surprised by the number of my neighbors who turn their leisure to some aspect of nature conservation, many of them in support of the activities of the Natural Resources Trust at Sheep Pasture. Conservation was an unlikely pastime when the factory bell woke people up at dawn and put them to bed, exhausted, at night. In *According to Season*, Mrs. Dana wrote: "That we know so little, as a people, of our birds, trees, rocks, and flowers, is not due, I think, so much to any inborn lack of appreciation of the beautiful or interesting, as to the fact that we have been obliged to concentrate our energies in those directions which seemed to lead to some immediate material advantage." Elise, Olivia, Oliver, and Richard Ames grew up at Sheep Pasture possessing sufficient material comforts that they were able to indulge a youthful interest in natural history without thought of the daily necessities of life, and they were fortunate to have useful instruction with nurse Matilda Golden and coachman John Swift. Now, more and more North Eastoners have the leisure for nature study, and we enjoy the expertise of the naturalists of Easton's Natural Resources Trust.

What became of the Ames children of Sheep Pasture? Oliver was killed in action in France during the First World War, leaving behind a new wife and a daughter born just before his death. Richard enlisted as a private, but because of his fluency in French he was assigned to responsibilities in strategic intelligence. After the war he settled in Paris, where he became a friend and confidant of well-known musicians. He died in 1935 and was buried in the Musicians Corner of Paris's Pere Lachaise Cemetery. Olivia served as a nurse at Fort Devens, Massachusetts, during the Great Flu Epidemic of 1918; she contracted the disease and was sick for many months. She later married and lived a long life. Elise, the eldest child, married William Parker in 1917 and maintained her association with North Easton. She inherited Sheep Pasture at the death of her mother in 1945. The mansion and stable were demolished the following year, to reduce a tax burden. When Elise died in 1979, the majority of the land and the remaining buildings were bequeathed to the Natural Resources Trust of Easton, for the education and edification of young people. I like to imagine that it was the influence of nurse Matilda Golden and coachman John Swift, and their lessons in natural history, that influenced Elise Ames Parker to make the legacy that would ensure that future generations of children might enjoy the same intimacy with the natural world that she knew during her own youthful excursions along the path.

BROOK

AT THE DAM on Main Street, Oliver Ames and his entrepreneurial sons squeezed the last bit of useful energy out of Queset Brook. The factory there, called the Red Mill, is long gone, and in its place is a fine vista across rolling fields designed by Olmsted. From the rocky pool below the dam the stream flows southward at a leisurely pace through Sheep Pasture. About a quarter mile from the dam the path crosses the brook on a plank bridge, and there, where the water lies dark and deep in a languid pool, it is hard to imagine that this unimpressive stream could have been the driving engine of shovel making. The earliest recorded name for this section of the stream is Trout-Hole Brook. When the brook was harnessed at the Main Street dam, it was called the Mill River or Sawmill River, and that was probably the end of the trout. At some time in the early nineteenth century, the stream began to be called the Cowisset River, most likely by being mistaken for another nearby stream named for a tribe of Native Americans. This eventually became the lovely Queset, a name decidedly more charming than the mundane Mill.

The plank bridge is one of my favorite places to sit, legs dangling over the side, day or night, summer or winter. The water here barely moves, but spreads wide to tuck itself into stony banks where fiddlehead ferns sprout in spring, and, on the opposite shore, into the roots of overhanging trees where orioles sing in May. Turtles sun themselves on whatever perches protrude above the surface of the water. Mallards waddle in the mucky shallows. In the deep shadows under the trees bass and pickerel dart and play. This is Darwin's dream pond, a teeming arena of eat and be eaten, a place that speaks to our own protoplasmic origins, the pond water of our blood, ancient urgings toward feeding and reproduction. "Love, we are a small pond," says Maxine Kumin in one of her wonderful poems. It is a delicious metaphor: the pond as tender affection, touching skin, the gentle scratches that leave no scars, the mouths that gobble. "The blackest berries fatten over the pond of our being," she writes, exuberantly. Her poem reminds us: We need to keep in touch with those things—the algae scum, the fiddleheads, the ducks, the turtles, the gush of life—

lest we forget what touch and sight and sound and scent are all about, and why the pond and river are important.

If I lie prone on the plank bridge, my face is only a foot or two from the surface of the water. Water striders and whirligigs skitter in late spring, back swimmers in the fall. Dragonflies dominate the summer pond, darting with helicopter versatility, establishing territories or seeking mates. Prone on the bridge, I watch male and female dragonflies curl their bodies into valentine-shaped embraces, coitus on the wing. Sometimes a mating pair stays locked together for an unseemly long time. They zip about in tandem, the male still grasping the female's neck while she lays her eggs. Like mating dragonflies, our drawing toward each other was cradled in the pond, nurtured on the tangled bank, perfected in the same urgency of seek and join that causes these agile fliers to bend their bodies into a heart-shaped kiss. *Love, we are a small pond.* The pond is more than a poet's metaphor for our lives; our lives are steeped in it.

WE LIVE ON the Water Planet. Nothing confirms this so vividly as a daytime crossing of the Atlantic Ocean at 30,000 feet. Viewed from an airplane in midocean, the sea stretches in every direction to the far horizons, apparently shrinkwrapping the planet in liquid H_2O. Photographs of the Earth from space show almost three-quarters of the surface covered with water or sea ice. Of the one-quarter of the surface that is dry land, a significant part is capped with ice, mostly in Greenland and Antarctica. Of the rest, more than half is under cloud at any given time—the visible signature of water in the atmosphere.

Liquid, solid, gas. What strange and wonderful stuff water is—almost as if our soppy planet were designed to be the cradle of life. Perhaps the most felicitous quality of water is that it is a liquid at moderate temperatures. Most other substances consisting of similarly small molecules—methane, ammonia, and hydrogen sulfide, for example—are gases. Liquid water is an excellent solvent which bathes living cells in nutrient-rich solutions, transports substances within cells, and helps flush away the toxic detritus of life. At the same time, water doesn't dissolve calcium phosphate, which is why our bones don't melt away. Of all liquids, water has one of the highest surface tensions, which allows capillary action to lift it up through the fibers of plants, from moist roots to thirsty leaves.

Water is so uniquely favorable to life as we know it, it is hard to imagine life without it. Where did it come from, this planetary wrap of fluid, this liquidy bower? A standard story is that the heat of the young Earth drove hydrogen and oxygen out of chemical combination in minerals like mica, and these atoms then combined to form water. Four billion years ago the planet was mostly molten, heated by radioactivity and the violence of its formation—a vast spherical volcano—and the newly formed water bubbled up out of the fiery depths as steam. Later, as the planet cooled, the Earth's shroud of gaseous moisture precipitated as rain, which collected in the broad, deep hollows of the newly formed crust to form the first oceans. "Our oceans were once our rocks," the chemist P. W. Atkins says of this scenario. Another possibility is that the water was already there in the gassy nebula out of which the solar system formed. A team of astronomers using the Infrared Space Observatory have observed what appears to be a massive water generator in a gas cloud in the constellation Orion, 1,500 light-years away. The Earth-orbiting telescope picked up the unmistakable spectral signature of water molecules, the largest concentration of water ever seen outside of our solar system. Like most interstellar gas clouds, the target nebula in Orion is mostly hydrogen, but it also contains free oxygen. A hot young star embedded in the cloud spews off powerful shock waves that pummel and heat the gas, causing oxygen to combine with hydrogen, creating enough water every single day to fill the Earth's oceans sixty times over. Eventually, the water vapor in the nebula will cool and freeze into small particles of ice, and particles such as these may have been present within the dusty cloud out of which our solar system formed, 5 billion years ago.

The Greek philosopher Thales of Miletus—who might reasonably be called the first scientist—believed water was the original substance of the universe. He observed that the seeds and nutriments of everything are moist, and from this he deduced that everything comes from water. Thales was certainly right about the primacy of water for life, but he was wrong to believe that water is the original substance. We now know that water is composed of two more basic substances—hydrogen and oxygen—and we have a pretty good idea where those elements came from. The hydrogen was apparently created in the Big Bang, 15 billion years ago; the oxygen was forged in the violence of exploding stars that lived and died before the Earth was born. What remains to be discovered is how and where the hydrogen and oxygen were forced together to create the quintillions of tons of life-giving water that cover the surface of our planet like a silvery sheath—and that purl in languorous eddies under the plank bridge across Queset Brook.

OLMSTED'S FIRST sketches for Sheep Pasture show the Queset Brook sculpted into curvaceous lagoons looped by carriage roads. This massive restructuring never happened, although modest changes to the stream were made. Just to the south of the bridge is what was once called "the Girl's Swimming Hole," where presumably Elise and Olivia took their dips on warm summer days, and, farther south near the woods, "the Boy's Swimming Hole," where Richard and Oliver bathed. I take my information from a watercolor map of North Easton painted many years later by Elise for her own sons. It is reproduced in an enchanting little book called *Growing Up at Sheep Pasture*, by Easton historian Hazel Varella, based on interviews with Elise, then Elise Ames Parker, in 1976. Water figures prominently on the map, as it must have figured prominently in the imaginations of the children. To the west of town on the map is Flyaway Pond (which lived up to its name in the great dam break of 1968), feeding a series of smaller ponds approaching the shops and mills of North Easton village. Across Main Street from Sheep Pasture, on Uncle John Ames's estate, is the romantically named Langwater Pond (or simply Fred's Pond to villagers, after John's father, Frederick Lothrop, who built the estate), with its graceful arched bridge designed by Olmsted. The Queset Brook makes its way through both estates, north to south, and is marked on the map along the way with remembered landmarks: "Oak Woods," "Goldfish," "Wood Ducks," "Wild Garden," "Lightning Tree," and of course "Girl's Swimming Hole" and "Boy's Swimming Hole." These latter pools have silted up since Elise and her siblings splashed in their waters, but the map suggests remembered warm-weather idylls.

Nineteenth-century children's literature, such as the books about Horatio Alger or Oliver Optic, generally offered moral tales of pluck, hard work, and obstacles overcome that led to commercial success, for which the first generations of Ameses, old Oliver and his sons, might stand as models. As the Sheep Pasture children were growing up, at the turn of the twentieth century, a different sort of literature was on offer. Perhaps the most popular fictional models for children at that time were the settled, middle-class Bobbsey Twins, fictional offspring of a wealthy lumber dealer in "Lakeport," created by author Lilian Garis in 1904. As historian Peter Schmitt has described in *Back to Nature: The Arcadian Myth in Urban America*, "The Bobbsey Twins knew nothing of pluck and peril. . . . Their values were those of affluent Americans who had time and money to make a virtue of outdoor life." The most pressing question for the young heroes and heroines of the Bobbsey Twins and similar books was "where shall we go for the next outing—because we must get into the woods somehow, and live close to Nature for a spell." The moral agenda of these books, Schmitt suggests, was to keep children from growing up too fast. The proper anecdote for premature urban precocity was a carefully supervised exposure to the tamed out-of-doors, and for one group of four young children the carefully tamed Sheep Pasture was just the ticket. The lives of Elise, Olivia, Oliver, and Richard echoed those of the Bobbsey Twins. Elise Ames Parker's watercolor map suggests little adventures along the Queset Brook: "Encounter in Checkerberry Wood," "Picnic at Twin Oaks," "Journey to Lincoln Spring." Nurse Matilda Golden would have been along on these outings to teach the wildflowers, coachman John Swift would be waiting at the end of the adventure with birdlore and snacks, and another family employee, Bunny Woods, taught the children how to catch snakes and put them in bottles.

The popularity of the Bobbsey Twins books endured into my own childhood, which was of sufficient affluence and proximity to tamed nature to provide similar Arcadian adventures. Adjacent to my home in suburban Chattanooga, Tennessee, were woods, fields, streams, ponds, and drainage ditches, and these were the venues for my play. We built dams and bridges, tree forts, and hideouts. Our companions were frogs, salamanders, crayfish, turtles, and snakes, fun to catch, keep, or simply use to scare away any girl who dared intrude upon our manly pleasures. My wife grew up in urban Brooklyn, the daughter of a fireman, but she also read the Bobbsey Twins and had her own Arcadian adventures in New York's neighborhood parks, legacies of Olmsted's era. She claims to remember individually every tree along her street, especially the "pollynose trees" (maples) and "itchyball trees" (sycamores). Several times a year she would make the longer journey to Brooklyn's Prospect Park, Olmsted's own masterful creation. In typically romantic fashion, the architect had given the park's features Bobbsey-Twinish names: Long Meadow, Lookout Hill, Breeze Hill, and the Nethermead, this last an archaic British construction for "lower meadow." Prospect Park had a zoo and botanical garden where children might observe for free exotic bits of Africa, Asia, and the Amazon.

Those excursions of our youth, in suburban countryside or urban parks, loom large in our memories; they gave us contact with the natural world that we might not otherwise have had. Whether or not those woodland frolics kept us from early maturity is debatable; they certainly gave us a sense of the organic unity of life. Ours was the last generation who read the Bobbsey Twins as a matter of course. My own children would have thought those books a bit unsophisticated, but nevertheless acted out their own Arcadian adventures with the nearby Ames estates to roam. When I accompanied the kids, I played the role of Matilda Golden, John Swift, and Bunny Woods rolled into one. We chased snakes, collected wildflowers, learned the name of a bird or two. In other words, a Bobbsey-Twinsish "back to nature" way of life survived for a century, more or less unchanged, and thanks to the generosity of the Ames family, it was not necessary for an Easton child to possess a private fortune to have the run of Arcadia.

IN WINTER we skated on Fred's Pond. The brook near the plank bridge seldom freezes, but above the "Red Mill" dam at Main Street the water freezes to a thick black depth. We loved to walk the ice at night, the glistening sheet resounding with an ominous rumble. When the ice was free of snow we skated, in darkness, trusting the blackness of the solid water beneath our feet.

By the standard of other substances, the properties of ice are bizarre, yet ice is so perfectly suited to our purpose that if it didn't exist we should have to invent it. With few exceptions, the solid phase of matter is more dense than the liquid phase; water, alone among common substances, violates the rule. As water begins to cool, it first contracts and becomes more dense, in the typical way. But about four degrees above the freezing point, something peculiar happens. It ceases to contract and begins expanding, becoming less dense. At the freezing point the expansion is abrupt and drastic. As water turns to ice, it adds about one-eleventh to its liquid volume. This extraordinary fact is portentous with meaning. It means that ice floats on liquid water, rather than sinks. It means that ponds freeze from the top down, rather than from the bottom up. It means that aquatic life in freezing climes can survive the winter in unfrozen water and mud under the ice.

Water in all of its states is essential to our welfare, including the water vapor in the air. Old Oliver Ames might have looked up at the clouds that billow in a summer sky and seen the power to hammer shovels from bars of iron. You and I might see the clouds as a source of beauty, or, if spilling rain, a nuisance. But the clouds are not just water droplets in transit from sea to mill pond, part of the water cycle. They are also subtle regulators of Earth's climate, primarily in two ways: They reflect back into space incoming radiation from the Sun, making the planet colder, and they keep surface heat from escaping to space (the greenhouse effect), warming the planet. Climatologists do not yet fully understand the bottom-line effect of clouds on climate (would increased cloud cover make us cooler or warmer?), which is one reason why predictions of greenhouse warming are so uncertain. It may be that Earth's average amount of cloud cover is self-regulating. It is also possible that a change in any of several variables—the overall balance of solar energy and its distribution by winds and ocean currents, or artificial greenhouse warming—might yield a more completely cloud-shrouded planet.

If the Earth were wreathed in cloud, like our sister planet Venus, no one would see a starry night, the changing phases of the Moon, the weaving dance of planets, or the geometric beauty of a solar or lunar eclipse. These are the very things that excited the scientific instinct among our ancestors. Historians of science Giorgio de Santillana and Hertha von Dechend contend that all the great myths of the world had their origin in the regular behavior of celestial bodies. Other commentators have stressed the connection between the heavens and the development of scientific thought. The stars might seem improbable objects to have aroused such curiosity since the human body is closer at hand and a more obvious candidate for systematic investigation. But astronomy advanced as a science before medicine, and early practitioners of medicine turned to the stars for signs and omens. The reason is clear: The regular motions of the heavens lend themselves to mathematical description. Beyond the apparent chaos of terrestrial experience, the stars proclaim the rule of law. Science was the child of the uncloudy sky.

On a cloud-shrouded Earth, the rise of the human species to civilization would almost certainly have been delayed but not forestalled forever. Sooner or later the inhabitants of a cloud-covered planet would advance to the point of finding ways to lift themselves above the clouds. They would get their first view of the universe beyond the clouds—the beckoning stars, the Milky Way, the luminous orb of the Sun, the changing Moon, planets and comets, solar and lunar eclipses—rhythms unhidden, the music of the spheres revealed, the rule of mathematical law, so laboriously learned in the terrestrial environment, in the heavens made crystal clear. "If the stars should appear one night in a thousand years," wrote Emerson, "how men would believe and adore, and preserve for many generations the remembrance of the city of God which had been shown. But every night come out these envoys of beauty, and light the universe with their admonishing smile." New England is certainly not clear every night, but when the stars are visible, I love to sit on the plank bridge across the Queset Brook, in a shower of starlight, listening to the barely audible slip of water beneath the bridge, seeing the light of those other Suns mirrored in the dark surface of the stream, reflecting upon the never-ceasing circularities of that most remarkable of substances, an oxygen atom holding hands with two hydrogen atoms, fountain of life, balm of Earth.

WHEN MY CHILDREN were young, we often played together on the plank bridge or along the banks of village streams, and no doubt the kids spent many more hours there on their own or in the company of friends. It was not, I think, time wasted. On the cover of Hazel Varella's *Growing Up at Sheep Pasture* is reproduced a bookplate used by Elise, Olivia, Oliver, and Richard Ames, designed by their kindergarten teacher Miss King. It shows the four children holding hands under a tree. There is an owl in the tree, representing wisdom, and an open book with the tried-and-true Victorian proverb "As the twig is bent, so the tree's inclined." For generation after generation, adults have bent children's twigs, or *tried* to bend them, and nature has figured strongly in their efforts. The Bobbsey Twins were part of the strategy, along with the Rover Boys, Bunny Brown and his sister Sue, and an army of other early-twentieth-century fictional tots tramping about in the near wilderness. The Boy Scouts and Campfire Girls offered real, as opposed to literary, exposure to woodcrafts, all of which was supposed to instill sturdy, self-reliant virtues. Growing up in Brooklyn, my wife moved from the Bobbsey Twins to Jack London's *Call of the Wild* and Gene Stratton Porter's *Girl of the Limberlost*. The very constancy of the notion that children should be exposed to nature suggests that there may be something to the sentiment. It is always good to know where we've come from, and if there is a single valuable lesson to be learned from nature it is that the universe is all of a piece.

My own kids were lucky to have had a year of schooling in London, England, and another year in Ireland where an important part of the curriculum was drawing from nature. Teachers took the students to the park or seashore to sketch what they found: bugs, leaves, blades of grass, shells, stones. The emphasis was not on art but on observation; not on self-expression but on faithful representation. The children were asked to look, see, and record what they saw. In London we lived near the British Museum of Natural History, a vast Victorian storehouse of natural diversity: stuffed animals by the thousands, glass cases full of glistening beetles and gaudy butterflies, room after room of dinosaur bones, rocks, gems, and fossils. Schoolteachers encouraged the children to go there on Saturdays. For the deposit of a large English penny (this was before decimalization), they were given a folding canvas stool, a drawing board, paper, and a fistful of colored pencils. Off they went into the depths of that cavernous building to sketch hummingbirds and pterodactyls.

My friend artist Clare Walker Leslie has devoted her life to encouraging American children (and adults) to draw from nature, but hers is a lonely mission. Not once in American schools were my children asked to draw from nature. They had art classes, yes, and good ones. They sketched sneakers, bottles, and bowls of bananas. In biology lab they drew what they saw under the microscope. But no teacher took them into a natural environment with pencil and paper. They were never asked to sit and sketch a mushroom in the woods or a spiderweb by the stream. The Natural Resources Trust of Easton runs a nature study program at Sheep Pasture for local schoolchildren. As I walk the path, I occasionally see groups of children clutching notebooks and observing stream life, birds, insects, or plants. But even this activity seems to have diminished when the grammar and middle schools in the village center were closed and consolidated with the big new schools (named for Olmsted and Richardson) at the edge of town.

The British and Irish emphasis on drawing from nature (which has lessened in those places, too) helped develop powers of observation and reinforced curiosity about the natural world. A child who has watched whirligigs and water striders on the surface of a stream will appreciate the importance of clean water. A child who has observed the clouds, their heapings and tumblings, their dark massings and silver linings, will be better prepared to understand the relationship between cloud cover and greenhouse warming. A child who has considered the beauty of a heron rising from the pond and the cunning of the spider's web will be more inclined to appreciate the importance of preserving biodiversity. With luck, nature study will also encourage curiosity about the hidden architecture of nature that only science can reveal—the properties of atoms, for example, that allow oxygen and hydrogen to combine to make water, the basis of life, and in its liquid, solid, and gas phases, the regulator of Earth's climate.

Scientific information and scientific method are important and must be taught in school, of course, especially in the upper grades. But more fundamentally, science is a set of attitudes about the world. It is respect for the evidence of the senses: seeing things as they are, and not as we wish them to be. It is conviction that the world is ruled by something more than chance and whim, and a confidence that the human mind can make some sense of nature's complexity. And science—almost paradoxically—is humility in the face of nature's complexity. One does not readily learn these attitudes in a classroom, but they can be learned by observing nature. Without these attitudes, children will grow up to be distrustful of scientific information, skeptical of scientific method, and poised to stumble willy-nilly into an uncertain environmental future. The huge problems we face as a species can only be solved with reliable knowledge of the way things work, and with a knowledge of nature's organic wholeness that only nature study can instill.

FROM WHERE THE PLANK BRIDGE crosses Queset Brook, the stream makes its way into the woods, to reappear again in South Easton on its course to Narragansett Bay. There it will debouch its waters back to where they came from: sea level, energy ground-zero, the water cycle completed. Every water molecule in the brook hankers for home, and in making its way back to the sea yields gravitational energy, potentially for human benefit. The Sheep Pasture estate was built with money squeezed from water by the older generations of Ameses. Now the stream and its environs have been harnessed for another use, the education of children, and the young naturalists on the staff of the Natural Resources Trust do their best to counter the lure of the video monitor. More important than the educational programs is the place itself, the fields, woods, and streams, open and available for play. As the water slips under the bridge, it whispers a promise that all of us would do well to hear, of a world contrived as a nurturing bower— clean water, clean air, green fields and forests, abundant wildlife—for as many of the planet's inhabitants as possible. *As the twig is bent . . .*

OPEN FIELDS

ON APRIL 11, 1779, Gilbert White of England's Selborne village wrote in his journal: "Ivy-berries ripe; the birds eat them, & stain the walks with their dung." It was the same day Oliver Ames, founder of the Ames Shovel Company, was born in West Bridgewater, Massachusetts. The United States was young and full of promise, Selborne already hoary with age. White was curate of the tiny village nestled in a quiet dale about forty miles southwest of London. Selborne today is a place of pilgrimage for many who love the natural world. White, who might fairly be called the first naturalist, wrote *The Natural History of Selborne*, a little book that remains in print today and is the model upon which I have based my thirty-seven years of study along the path. According to White, any place can be every place if attended to with care. Within the confines of his tiny village he found sufficient things to occupy his interest for a lifetime. He knew that the particular contains the all.

I once had the very great pleasure of visiting Selborne. The village has changed little since White's time. I walked through the rooms of the beautiful old house on the village green where White lived from 1730 until his death in 1793. I visited the garden behind the house, where he tended vegetables, fruit, and flowers, and where Timothy the tortoise presided. I tramped the beechwood "hanger" above the village, where White sometimes heard the nightingales' song, and the path that follows the gentle brook that flows from Selborne Green to the ruins of Selborne Priory. British conservation groups, particularly the National Trust, have carefully preserved White's village and the surrounding lands. To visit Selborne is to step back in history, to a simpler time, when the natural environment still pressed close upon consciousness and the call of the cuckoo served to punctuate the year. The landscape of Selborne is not dissimilar to the landscape of my path. After all, it was the late-eighteenth-century English countryside that Olmsted took as his model for American public parks and private estates.

For twenty-five years White kept a nature diary. In it he recorded such mundane facts as temperatures and barometer readings, measures of wind speed and rain. He also entered brief observations of anything in nature that attracted his attention. Those observations, each slight in itself, become in the accumulation a kind of poetry:

> Jan. 15. Hailstones in the night.
>
> Jan. 25. Snow gone. The wryneck pipes.
>
> Feb. 17. Partridges are paired.
>
> Feb. 21. Ashed the two meadows.
>
> Mar. 14. Daffodil blows.

On the title page of his journal for 1768, White copied this motto, "I solitary court the inspiring breeze and meditate the book of Nature ever open," and in the pages that follow he maintained an accurate reading of Nature's book. What is missing from White's journals is almost any mention of the greater world beyond Selborne village, even though he lived in a time of social upheaval. The Industrial Revolution was beginning in England and would soon be transplanted to America by men such as Oliver Ames; of this, there is no hint, except for one brief reference to mechanical weavers at Alton, a town not far from Selborne. The American colonies asserted their independence during the time White was writing; he makes but a solitary note of the victory of General Cornwallis in North Carolina. What he considers instead are the ordinary events within a few square miles of English countryside:

Apr. 10. Therm. 72!!! Prodigious heat: clouds of dust.

Apr. 18. A nightingale sings in my fields. Young rooks.

Apr. 20. Some whistling plovers in the meadows toward the forest.

Apr. 27. Many swallows. Strong Aurora!!!

White was a meticulous observer. Not much happened in his village that escaped his eye. The reason his book and journals remain in print today, two centuries on, is the way he understood the interrelationship of all living things. He was an ecologist before ecology had a name. He recognized, for example, that lowly earthworms are essential to the fertility of the soil. (They aerate the soil and hasten decomposition.) He observed cattle standing in ponds on warm afternoons and understood their importance to life in the pond. (Insects in cow dung are food for the fish.) Nature is the "great economist" he said, who shares around its resources. He taught us to understand *that we are part of the web*. There is no nature for Gilbert White that does not include human nature.

One human character in *The Natural History of Selborne* gets a chapter all to himself: the bee boy. The mentally disadvantaged young man, who was lean and of "cadaverous" complexion, had one consuming interest: honeybees, bumblebees, and wasps. These insects were "his food, his amusement, his sole object." The bee boy sought out bees wherever he could find them, never giving a thought to their stings. He grasped them bare-handed, plucked out their stingers, and sucked their bodies for their honey. Sometimes he would stuff swarms of bees beneath his shirt, against his bare skin, and take them to his home. To the annoyance of local beekeepers, the bee boy would slip into their gardens, rap with his fingers on the hives, and grasp the bees as they came out. Sometimes he tipped over hives to get the honey, of which he was passionately fond. Wherever men made mead from honey, he hung about, begging for a drink of what he called "bee wine." As he ran about the village, he made a humming noise with his lips, resembling the buzzing of bees. And in the winter, when bees kept to their hives, the bee boy dozed away his time by the fireside of his father's house, in a torpid state, waiting until the spring and summer when once again he hummed about the fields and gardens in search of his prey.

A present-day psychologist might find a plausible explanation for the bee boy's curious fixation. But who will fault the boy his passion, equaled in all that village perhaps only by Gilbert White himself? The boy focused his fierce attention upon a single object; White disbursed his attention widely, anointing every stone, bird, insect, and blossom with his study. Most of us live out our lives somewhere between the bee boy's obsessive single-mindedness and White's encyclopedic interest. When we read of

White cutting open snakes, birds, and hedgehog dung to see what is inside, we flinch and wonder if curiosity really requires such attention to detail. But we also pity the bee boy for his universe of lost appreciations: butterflies unseen and nightingales unheard. He carried to a pathological extreme a natural tendency to lose oneself in the particulars of nature.

In a short poem called "Beginning My Studies," Walt Whitman spoke of the temptation to never go beyond the first object of his study—"the least insect or animal"—and submerge himself in its intricacies, to "stop and loiter all the time to sing it in ecstatic songs." Whitman knew that there is a sense in which the least thing contains the all. "The nearest gnat is an explanation," he proclaimed in "Song of Myself." If the plan of the world is simple and universal, as every scientist assumes, it will show itself as well in a single bee as in a landscape full of plants and animals, and as fully in a single local landscape—the open fields beside my path, perhaps—as in a sprawling universe of galaxies.

This is why the mad bee boy kneeling at the hive is as close to divinity as the sane naturalist who takes the whole village as his ken. Our understanding of the world is advanced by both the Charles Darwins and the Gregor Mendels: the generalists who range across continents and oceans of experience, and the specialists who are content to tease out the laws of genetics from a single patch of peas. I do my own share of ranging, I suppose, mostly in books, acquiring the knowledge—general and particular—that illuminates the landscape of my path. But I am also content to sit for hours in a single meadow, observing the commerce of insects and wildflowers or waiting for a glimpse of the vee-vested meadowlark that teases me from its hiding place with its three-note trill. I am guided by Gilbert White's and the bee boy's implicit faith that the universe reveals its secrets at every level of complexity.

FROM THE PLACE where the plank bridge crosses Queset Brook, a long rolling meadow reaches up to the site of the Sheep Pasture mansion. This is the bonny lea where the Ames children frolicked on the last warm days of late autumn, picking wildflowers and chasing butterflies, watched over by Matilda Golden, their nurse. And here too, not long ago, I walked with my students and spied the last monarch butterfly of the season. We watched it flit across the meadow, a matchstick with wings. No, not wings—gaudy flags, orange and black. Semaphoring southward, Mexico bound. We stood and watched in rapt contemplation until it disappeared beyond the brook. *Adios, amigo. Hasta la vista.* See you next year, or rather your nieces and nephews of the next generations.

If any creature embodies within itself the secrets of cosmic complexity, it is the monarch butterfly. The monarchs arrive in our neighborhood in late spring and early summer. They mate and lay eggs on the leaves of milkweed, and milkweed only. The eggs hatch a few days later, and the larvae feed on the milkweed leaves. The leaves contain poisons, but these seem not to bother the monarch caterpillars. They store the toxins away in their bodies, which deters birds from eating them. The yellow-, white-, and black-striped caterpillars molt a few times, then pupate. Inside a chrysalis the larva rearranges its molecules to make a butterfly, preserving the toxins intact. The caterpillar's six stumpy front feet are turned into the butterfly's slender legs. Four bright wings develop, as do reproductive organs. The caterpillar's chewing mouth parts become adapted for sucking. A few weeks later the adult bursts forth with its glorious flags and toxic flesh. Few birds are dumb enough to take a bite.

There may be three or four monarch broods in a summer. Then, in September or early October, the last generation of adults heads south. No one knew where they went until 1975, when Canadian zoologist Fred Urquhart discovered hoards of monarchs roosting in a few patches of mountainous forest of central Mexico. Adult butterflies from all over eastern North America congregate in these clumps of trees and rest through the winter, as thick as leaves. The migration of the monarchs is one of the epic journeys of the animal kingdom. They fly by day and rest by night, singly or in groups. Millions finally arrive in Mexico. My students and I talked of this as we watched the last monarch of autumn drift out of the meadow, beyond the brook, sunward, southward, called by some mysterious inner voice to a clump of fir trees thousands of miles away. Why didn't earlier broods of summer hear the call? A seasonal trigger, certainly. A chill in the air. The waning sun. And then the voice, speaking a language whose words are chemical, compelling, irresistible: "Take wing. Follow me." Down along the spine of the Appalachians, across the flat croplands of the Gulf coast, into the mountains of Mexico.

In the spring the overwintering monarchs leave their Mexican roosts and head north, breeding and dying along the way, so that it is a new generation that returns to our northern meadows. There will be no Mexican veterans to lead the next southward migration. How do they do it? How does a butterfly that has never made the journey know when to depart and where to go? That tiny pinch of toxic flesh with bright postage-stamp wings? This much seems certain: The call, the map, the skills for navigation were there in the egg, somehow encoded in the monarch's DNA. Not, of course, in the way maps and directions are stored in the memory of an automobile or airplane's navigation system. It is far more subtle than that. What the DNA encodes for are proteins. The proteins are a language of geometry that speaks through the monarch's nervous system: *"Ven, sigueme!* Come, follow me." No one knows how the language evolved; nor does anyone yet understand the vocabulary or syntax. Surprisingly, a butterfly's DNA is more voluminous than our own, forty times more voluminous! Most of the butterfly's DNA is probably junk—long strands of useless garble, the derelict residue of the butterfly's evolutionary history. Other parts code for the successive body designs of an insect that goes through metamorphosis. Somewhere in that encyclopedia of information is the invitation to a journey whose origin is lost in the depths of time.

The migration of the monarchs is just one of many miracles of life that await explication by science. Make no mistake; someday the miracle will be unraveled, the DNA sequenced, the proteins decoded, the subtle environmental signals and their receptors mapped—if only the monarch survives long enough for its secrets to be exposed. Its survival, of course, is in jeopardy. Those patches of Mexican forest where the monarchs roost are threatened by logging. Only a few sanctuaries are protected, and those precariously. The noxious chemicals we put into the air and the soil here at home don't help. Our milkweed meadows are giving way to suburban housing and industrial complexes. That we might willfully delete from the universe so wondrous a thing as the monarch migration is a possibility too sad to be reckoned. In *The Magic Mountain*, the novelist Thomas Mann defines life this way: "It was a secret and ardent stirring in the frozen chastity of the universal; it was a stolen and voluptuous impurity of sucking and secreting; an exhalation of carbonic gas and material impurities of mysterious origin and composition." Gushy, over-the-top prose, yes, but why not? Life *is* gushy and over-the-top. What could be more over-the-top than those clouds of colorful fliers—pumpkin, salmon, and flame—beating their way from our northern meadows to havens of fir trees in Mexico?

A FEW YEARS AGO, it was my privilege to visit the Mexican monarch sanctuaries with a group of fellow naturalists. Getting there wasn't easy. A six-hour drive from Mexico City over sometimes frightening mountain roads, with an overnight stay along the way in Zitacuaro. A mile-long climb by foot along a rugged trail deep in volcanic dust to 10,500 feet, then a drop into a canyon forested with firs. We had seen a hint of what we were looking for as we passed through the mountain village of Angangueo on our way to the refuge: hundreds of bright orange monarch butterflies dancing in the air, casting fluttering shadows on walls, faces, pavement. The villagers went about their business as if these colorful presences were the most normal thing in the world. But nothing could have prepared us for what we saw when we finally got to the tiny patch of forest where 20 million monarchs covered the trees more thickly than leaves. When occasionally sunlight flooded through a break in the clouds, the butterflies took flight in their billowing millions. If you listened carefully, you could hear the soft rush of their wings. As I watched, wide-eyed and dropped-jawed, it occurred to me that among this teeming mass of monarchs might be one of those I had watched the previous fall in the Sheep Pasture meadows.

To certain people, these few patches of Mexican forest, winter home to the monarchs, are merely stands of lucrative timber to be harvested at will. In fact, the winter roosts are key to the monarch's survival—what monarch expert Lincoln Brower calls the insect's "Achilles heel." Not long after the discovery of the monarch roosting places, conservationists realized that the trees must be saved if the monarch is to survive. Led by Mexican poet Homero Aridjis, they pressured the Mexican government into establishing small sanctuaries enclosing the principal areas of congregation and holding the logging interests at bay. But now the butterflies face another threat, and I was part of it as I scrambled under the fluttering fir trees with my camera: ecotourism. Today, growing numbers of sightseers troop through the forest to see one of the great wonders of the natural world. Can the refuges sustain such popularity?

As I made the long climb along the dusty trail that led to the clustered butterflies, I observed the other people who made the trek. A few were *Yanque* tourists like myself, with our fanny packs, Vibram soles, expensive cameras, and binoculars. But the great majority of folks along the trail were Mexican,

and, as far as I could see, they were not the sort of affluent, middle-class sightseers you'd meet in Yellow-
stone or Yosemite. They were people of all ages—old
men and women, children, and everyone in between—apparently country people or city dwellers
who had not yet lost their country ways. Like the rest
of us, they struggled along the steep trail, choking on
dust. The only place I have seen similar assemblies of
trekkers—the very old and the very young making a
difficult climb—is on the holy mountains of western
Ireland on the annual days of religious pilgrimage.

As we reached the tiny clump of trees festooned
with butterflies as thick as jungle foliage, we Yanks
buzzed about, snapping pics, taking notes, storing
up impressions with which to later regale our
friends back home. The Mexicans by and large sat
silently in the forest, kids in laps, eyes somberly
fixed on the massed monarchs. It was difficult to
read their emotions, but I believe that many of the
Mexican visitors to the Chincua Monarch Sanctuary were driven by the same urge that might have
led them on another weekend to the Virgin's shrine
at Guadalupe: a sense of the holy. And further, I
think that unless those of us with our Vibram soles
and fanny packs can reclaim a sense of the holy, the
monarchs will remain endangered. By holy, I refer
to whatever it is in the ceaselessly spinning DNA
and chemical machinery that causes a creeping
caterpillar to rearrange its molecules into a winged
angel, and sends that angel fluttering across a conti-

nent to a patch of fir trees in Mexico it has never seen before.

The philosopher William James said, "At bottom the whole concern of both morality and religion is with the manner of our acceptance of the universe." What I think I saw on the faces of the Mexican visitors to the Chincua Sanctuary was a dignified and unquestioning *acceptance*, an understanding that what they saw was natural and right and utterly essential to the completeness of creation. The poet E. E. Cummings wrote of acceptance "for everything which is natural which is infinite which is yes." Science and politics alone will not save the monarchs, any more than they will save other threatened species and habitats, any more than they will save the monarch meadows and milkweed stands along my path. What is required is something that we have mostly lost in the high-tech, high-velocity, virtual world of the developed countries: a deeply felt, unintellectualized, instinctive "yes"—a sense that behind the gaudy delight of 20 million butterflies hanging on fir trees, there is a natural and infinte power that binds all life in a holy web.

ON WARM AUTUMN AFTERNOONS I have often sprawled in the monarch meadow that rises from the Queset Brook. Yellow goldenrods and purple asters fall away to the purling stream. Insects patrol the blossoms and dance in air. Birds sing.

You'd never guess in this magical place that else-
where in the world people are worried about environmental degradation. But, of course, the planet is
threatened by several sorts of environmental catastrophe, some of which might take place on a scale
that dwarfs the danger to the monarch's wintering
grounds. Technology is a mixed blessing. On the
one hand, it has helped to control diseases, enhance
communication, increase wealth, feed billions, and
even makes it possible for people such as myself to
follow the monarchs with some measure of con-
venience from New England meadows to Mexican
forests. On the other hand, technology supplies the
instruments to efficiently clear forested mountains
in the south and pave over milkweed meadows in
the north—or to devastate monarch populations
with chemical or genetically engineered pesticides.

Many of us remember the happy dream of the
1950s, when science promised to wipe out insect-
borne diseases with DDT, including especially the
worldwide killer malaria. Fogging machines went up
and down the streets and beaches of America spray-
ing insecticide into the air, with the promise of better health for all. In January 1958, a woman named
Olga Owens Huckins sent a portentous letter to the
Boston Herald, with a copy to Rachel Carson, a biologist and writer already famous for her books *The Sea
Around Us* and *The Edge of the Sea*. Huckins and her
husband owned a private bird sanctuary in Duxbury,
Massachusetts. As part of a mosquito control pro-

gram, and without permission, the state sprayed the property from the air with DDT, leaving dead song-birds in its wake. Huckins expressed her anger in the letter, and the rest, as they say, is history. Carson did her research, then wrote the book that many consider to be the most influential environmental tract ever written. *Silent Spring* soared onto the best-seller lists and was translated worldwide. Rachel Carson became the most famous woman scientist in the United States. And the environmental movement as we know it was born, perhaps just in time. In 1965 I saw the last bluebird along the path that I would see for thirty years.

Carson was subject to bitter attacks from the chemical industry and its allies in the federal government. As her biographer says, "She was questioning not only the indiscriminate use of poisons but the basic irresponsibility of an industrialized, technological society towards the natural world." Carson's views prevailed, and DDT was banned in the United States and in many other parts of the world. Today, bluebirds are back at Sheep Pasture. There are trails of bluebird boxes up and down the meadows and along the hedgerows, and every season more boxes are occupied by bluebird breeding pairs, fledging two, even three, nests of chicks.

The chief reason for our local bluebird renaissance is a fellow named Bob Benson, a big man with a big heart whom I often meet as I walk the path at almost any time of the day, but especially in the early morning. Bob minds the bluebirds, records their numbers, charts their growing population. He protects the boxes from occasional human vandals and marauding chipmunks. He cleans the boxes between broods and sometimes replaces a chick that has fallen prematurely from the nest. He does all of this with no other motive than love of the birds and the pleasure of being in the natural world. "If success were measured by how much money I have in the bank, the world would deem me a failure," Bob once told me, "but out here I'm the richest man in the world."

When Bob was a child, a bluebird family made its nest year after year in a hollow fence post near his home. He would sit for hours, rapt, watching the birds. Then the state sprayed for gypsy moths, and the bluebirds disappeared. He didn't see another bluebird for decades. Bob takes pride in the fact that it is the "little" people like himself, all over the country, who have brought the bluebirds back. "We don't often get a second chance," he says. He sometimes worries that he does *too much* for the birds. ("Next thing they'll be wanting towels and soap dishes," he jokes, as he fills a tray with water.) What's developing between Bob and the bluebirds is a kind of symbiosis—a relationship of mutual benefit—something quite common in the natural world. I think that what Bob is doing is good for the birds, good for Bob, and good for the rest of us. His is the spirit that will save the monarchs, too, if they can be saved.

Although bluebirds are again thriving at Sheep Pasture, malaria continues to be the world's biggest killer of children. Since the withdrawal of DDT for malaria control in South Africa, cases of the disease have quadrupled. Many public health officials now call for renewed use of DDT in certain malaria-ridden parts of the globe, at least until a vaccine or genetic fix comes along. Did *Silent Spring* save songbirds in New England and put babies at risk in Mozambique? Our technological interventions in nature usually involve a muddy moral arithmetic. Many people are vehemently opposed to genetically modified organisms, for example, but how should they feel about modifying the malaria parasite's genome in such a way as to disrupt its deadly ability to prey on the human immune system? Or engineering a malaria-resistant mosquito to replace natural populations? In the meantime we debate the ethics of DDT. Birds versus children: How do we balance our sometimes competing agendas? Albert Schweitzer said, "Man can hardly even recognize the devils of his own creation." We are not that good at recognizing the angels either.

MANIPULATING NATURE for our own sake is not a modern phenomenon. Any open field in a naturally forested clime is a technological artifact. New England's first fields—if that's what they could be called—were sporadically opened by Native Americans for planting corn. When colonial farmers arrived, they brought to this continent the permanent field systems of the Old World. Field boundaries in Europe, present and archaeological, inscribe a record of 3,000 years of human intervention in the landscape. The invention of crop rotation, and particularly the planting of clover for restoring nitrates to the soil, made it possible to keep fields continuously in use.

The meadows along my path have basked in sunlight for three centuries since the felling of the forests. A dozen parcels of agricultural land were assembled by Oliver Ames to create the grassy vista his family enjoyed from the terrace of the Sheep Pasture mansion. Since that time, the open fields have been maintained primarily for aesthetic, rather than agricultural, reasons.

It is widely acknowledged that a landscape of open fields, trees, and brooks is what humans consider most beautiful, perhaps because of our species's long childhood in the parklike savannas of East Africa. The biologist E. O. Wilson has suggested that whenever people are given a free choice, they move to open tree-studded land on prominences overlooking water; they are responding, he says, "to a deep genetic memory of mankind's optimal environment." This is exactly the sort of landscape that Olmsted created at Sheep Pasture. My walk to and from work is a kind of time travel back to the land of my most ancient forebears, who walked the grasslands of East Africa, hunkered in the shade of trees, gazed at their reflections in placid pools.

These open spaces along my path are utterly alien to New England's primeval forested landscape, but now after a few hundred years of "optimization" they seem as natural as—well, as natural as the much more ancient English countryside they were intended to emulate. The now-demolished Sheep Pasture mansion and its gatehouse (which still exists) were designed in English half-timbered style; it was up to Olmsted's team of designers to complete the illusion with grounds that might be situated in the English Midlands, at Stowe, perhaps, in Buckinghamshire, one of those informally landscaped estates in the Arcadian style that defined Olmsted's taste.

The owner of Stowe, Viscount Cobham, discarded the formal, rectilinear gardens of the continent for something more self-consciously "natural," with "Elysian Fields," temples, grottoes, and graciously curving shorelines, Disneyesque in an eighteenth-century sort of way, but catering to a deep-seated longing for the blending of the cultural and the wild. Part of the Stowe estate, the so-called Grecian Valley, was designed by Lancelot "Capability" Brown, the eighteenth-century founder of what we in America might call the Olmstedian style of landscape design. The Grecian Valley is the largest and least formal part of the gardens at Stowe and was meant to look untouched by human hands, although in fact the valley was dug by many laborers wielding shovels and bar-rows. The British landscape historian Christopher Hussey sees Whiggish politics at work at Stowe, ar-chitecture bringing the landscape "into harmony with the age's humanism, its faith in disciplined freedom, its respect for natural qualities, its belief in the indi-vidual, whether man or tree, and its hatred for tyranny whether in politics or plantations." Not bad qualities to aspire to then or now.

Given the sensibilities of those times, it may be that a village or two of poor tenants was displaced to create the expansive gardens at Stowe, but at least the property is now in the care of Britain's National Trust and open to all the people, a national treasure. Sheep Pasture, too, was once a pleasure garden for the rich; like Stowe, it is now in the public domain, and we are better for it. The poet James Thomson visited Stowe in 1734, and in his poem "The Sea-sons" describes "th' enchanted round I walk / The regulated wild." It hardly matters whether the rea-son we enjoy these Arcadian landscapes is genetic or cultural: they feed our hunger for the regulated wild and serve the monarchs and bluebirds, too.

WATER MEADOW

BEYOND THE PLANK bridge over Queset Brook, the path wends between rolling upland fields and a broad lowlying meadow that floods after any winter melt or major rain. Sometimes the path itself floods, and I must take off my shoes and wade through icy water. In spring, especially, the water meadow teems with life. Red-winged blackbirds take up residence in the young alder and willow trees that are colonizing the meadow. Canada geese and mallards, in faithful pairs, swim among emerald shoots of new grass. Then comes the magic day in April when the entire surface of the water meadow begins to sing, the choiring of spring peepers, a hallelujah chorus in celebration of new life. Of course, it's not new life at all. Life was there all along, buried in the frozen mud or in rock-hard seed cases on the branches of trees. Life doesn't come and go; life persists.

Often I have taken off my shoes, rolled up my pants, and waded out into the midst of all this animation. Every sense is stimulated by the brash exuberance of living things. And yet I know that the real exuberance I cannot sense at all. It occurs at the level of the molecules of life, a ceaseless chemical dynamic, spinning and weaving, capturing sunlight, respiring, copying, correcting. That fundamental chemical substrate is pretty much the same in every creature—frog, goose, red-winged blackbird, willow, grass, bacterium; behind the apparent complexity of life is an astonishing simplicity. In his autobiographical book *The Double Helix*, James Watson, the codiscoverer of the structure of DNA, tells how he came to think of a helix as the structure for that molecule. "The idea [of the helix] was so simple," he says, "that it had to be right." I think of Watson's words every time I look at computer-generated images of DNA, proteins, hormones, and the other molecules of life that are published with increasing frequency in scientific journals. These molecules are complex—they are made up of thousands, even billions, of atoms—and yet they are so simple and so beautiful that we look at them and *we know that they are right*. If I focus on the "atomic" dots of these computer images, I am bewildered by the astonishing complexity of detail, and I wonder that life exists at all. But when I focus on the overall patterns—the geometric forms and symmetries of the molecules—I am dazzled by an almost inevitable simplicity. Every molecule seems miraculously contrived for its task.

Some images of the molecules of life, as they are displayed on the color computer screen, resemble the gorgeous stained-glass windows and soaring architectural members of the Gothic cathedrals. A cross-section of the B DNA double helix, for example, bears a likeness to the magnificent rose window at Chartres. The webbed vaulting of the clathrin protein and the flying buttresses of the sugar-phosphate side chains of the DNA evoke the same sense of architectural déjà vu. No medieval architect could have raised more fitting structures. The medieval builders wanted the visible structures of the Gothic cathedral to reflect the invisible realities of the spirit world; the cathedral was conceived as an earthly image of the kingdom of God. Similarly, computer representations of the molecules of life are attempts to represent an unseen reality with visible images. And here, too, there is an almost mystical vision of a hidden harmony that has been established throughout the cosmos.

Behind the water meadow's rich diversity of life there is a molecular unity, so simple "it had to be right," as beautiful as a cathedral's rose window or flying arch. Abbot Suger of Saint-Denis, one of the greatest of the Gothic builders, hoped that his cathedral would reveal the divine harmony that reconciles all discord, and that it would inspire in those who beheld it a desire to establish that same harmony within the moral order. The molecules of life, revealed by science, achieve the same effect. They inspire a reverence for the invisible harmony—of form and function, of complexity and simplicity—that is the miracle of life. Here in the water meadow it burns like a ceaseless flame.

WE HAVE, it seems, a fierce attraction for spirits: auras, angels, poltergeists, disembodied souls, out-of-body experiences. Mostly, I think, we are drawn to these things because we intuit—correctly, it turns out—that there *must* be more to the world than meets the eye. We inherit the spirit world from a time when our ancestors huddled in dark shelters at night and let their imaginations draw up creatures more or less like ourselves although lacking corporeal substance. But why should we care about angels when the season's first blackbirds spread their red-shouldered wings? Why should we seek treasures in Heaven when year after year the fiddlehead ferns unfurl their silver croziers along the brook? Why should we look for out-of-body experiences when it is our *bodies* that connect us through the five open windows of our senses to the sights, sounds, tastes, smells, and tactile sensations of nature? If we want more than meets the eye, we should practice on this: the invisible flame of the DNA.

Even as I stand motionless and attentive at the

edge of the water meadow, a flurry of activity is going on in every cell of my body. Tiny protein-based "motors" crawl along the strands of DNA, transcribing the code into single-strand RNA molecules, which in turn provide the templates for building the many proteins that are my body's warp and weft. Other proteins help pack DNA neatly into the nuclei of cells and maintain the tidy chromosome structures. Still other protein-based "motors" are busily at work untying knots that form in DNA as it is unpacked in the nucleus of a cell and copied during cell division. Others are in charge of quality control, checking for accuracy and repairing errors. Working, spinning, weaving, winding, unwinding, patching, repairing—each cell is like a bustling factory of a thousand workers. A trillion cells in my body are humming with the business of life. And not just in my body. The frogs singing from their hiding places—their cells are in a flurry, too. The mallards paddle-wheeling through the flooded grass. The gelatinous scum of frog eggs at the water's edge. All of it invisibly astir. The more one thinks about it, the more unbelievable it sounds.

Oscar Wilde said, "The true mystery of the world is the visible, not the invisible." The smallest insect is more worthy of our astonishment than a thousand sprites or poltergeists. Scientists have now provided a complete transcription of the human genome, a listing of the tens of thousands of genes that are the plan of a human life, many of which we share with other creatures. They have found ingenious ways to manipulate the DNA—stretch it, snip it, splice it, watch the protein motors at work, measure their speed. What a thing it is to think of ourselves as manifestations of this invisible molecular machinery, ceaselessly animating the universe with sensation, emotion, intelligence. To say that it is all chemistry doesn't demean the dignity of life; rather, it suggests that the most elemental fabric of the world is charged with potentialities of a most spectacular sort. We have perhaps an infinite amount yet to learn about the molecular chemistry of life, but what we have already learned stands as one of the grandest and most dignified achievements of human curiosity. Forget all that other stuff—the spooks, the auras, the disembodied souls; *embodied* soul is what really matters. As I stand by the water meadow, I try to refocus my attention away from the ducks and geese and trees and frogs (and human observer), and attend instead to the thing I cannot see but know to be there, the endlessly active, architecturally simple unity of life—the meadow aflame, burning, burning.

IN MY MOTHER'S LIBRARY when I was growing up, there was a book by the naturalist writer Donald Culross Peattie titled *An Almanac for Moderns*, published in 1935, in the depths of the Great Depression. Peattie lived in rural Illinois at the time, and the devastation of the Dust Bowl was not far away. World War I was still fresh in memory, with its shattered landscapes and poisoned air. It was not a time when it was easy to be optimistic. The lofty moralizing of earlier nature writers like John Burroughs and John Muir no longer resonated with a generation who had seen "the trees blasted by the great guns and the birds feeding on men's eyes." The confident, cheery lessons of the Bobbsey Twins, which Peattie might have read as a child, must have seemed spoiled and hollow. Peattie, like Loren Eiseley and Lewis Thomas after him, looked skeptically at nature, not expecting edifying sermons in leaf and stone, and found a chastening silence. Yet he wrested from his observations of nature the will to go on, a point to life.

W. H. Auden said of Loren Eiseley that he was "a man unusually well trained in the habit of prayer, by which I mean the habit of listening." Peattie also knew how to listen. Listening—as Eiseley, Thomas, and Peattie listened—requires the courage to surrender traditional pieties and embark upon an immense journey into the lonely spaces between the galaxies and the atoms. From his closely observed acre of land in Illinois, Peattie listened and watched as the year passed, and turned his "habit of prayer" into a collection of 365 elegant reflections that wrestled with the meaning of life. The meaning he found in nature had something to do with beauty. Something to do with the gorgeous, prodigious throb and thrust of life. Something to do with being part of a continuity that was greater than himself.

"I say that it touches a man that his blood is sea water and his tears are salt, that the seed of his loins is scarcely different from the same cells in a seaweed, and that the stuff of his bones are coral made," he wrote in *An Almanac for Moderns*. He was immersed up to his neck—nay, to the top of his head—in the "essential and precious something that just divides the lowliest microorganism from the dust." He reveled in it, and he turned his experience into poetry. Peattie did not look for an incorruptible Heaven beyond the stars. Nature is miracle enough, he wrote, with all its imperfections.

In search of that "essential and precious something," I step barefoot into the cold waters of the water meadow in search of the source of the vernal song. Mud squishes up between my toes; a month earlier the meadow had been the sarcophagus of frogs. All winter long, the adult amphibians have hibernated in their subterranean tombs, waiting for the warmth of spring. Now they rise and lend their voices to the general celebration—birds, insects, a million peepers. I move among the grasses, the willows with their burning tips of redwings' epaulets. What other creatures harbor in the slime? What are these masses of gelatinous spheres? This scum of algae? Is this what Peattie called the "most unutterable thing" in evolution, "the terrible continuity and fluidity of protoplasm, the inexpressible forces of reproduction—not mystical human love, but the cold batrachian jelly by which we vertebrates are linked to things that creep and writhe and are blind yet breed and have being"? It can be a little frightening to attend to our kinship to the slime, but to do otherwise is to ignore the great polar extremities of existence: the individual and the collectivity, birth and death, generation and decay.

THE MID–NINETEENTH CENTURY was fossil time in science. As Oakes and Oliver Ames were manufacturing two-thirds of all the world's shovels, professional scientists and amateur naturalists were using some of those shovels to dig for fossils. Within a few decades in midcentury, tens of thousands of fossil animals and plants had been named and classified. Some were similar to living organisms. Others were strikingly different. Of course, fossils had been a source of speculation since antiquity; the rocks are replete with stony organisms. But the Industrial Revolution opened up rocks for examination as never before; mines, roads, railroads, and canals tore through the strata. Biographer Adrian Desmond says, "It was the equivalent of finding a new continent of creatures, underground." Perhaps never before in human history had so much *intractable* knowledge been acquired so quickly. Then, in 1859, Charles Darwin published his great book and turned the world upside down.

The unity of life by common descent over geologic time gave paleontologists the conceptual framework they needed to make sense of the fossils. Suddenly, everything fell into place; the scattered jigsaw pieces fit together to make a coherent picture. Darwin's champion Thomas Huxley had seen the implications of the fossils. He wrote, "To the very root and foundation of his nature man is one with the rest of the organic world."

Something similar is happening today, another flood of biological data, another hidden continent revealed, this time not in the rocks but in the cells of living organisms, those spiral staircases of chemical code (paired molecules of four kinds, A, C, G, and T, called bases) that are common to every living organism. Human DNA contains several billion base pairs, something like 30,000 genes coding for many tens of thousands of proteins. The sum of all the genes is an organism's genome. A fruit fly's genome is less than one-tenth the size of a human's. A bacterium's genome might be only one-thousandth as big as a human's. As the twenty-first century began, the genome of only one multicell organism, a tiny worm called *Caenorhabditis elegans*, had been fully sequenced. Today, the data bases are overflowing with As, Cs, Gs, and Ts. Every month a new creature yields its symphonic score.

Francis Collins, director of the National Human Genome Research Institute, has said, "The stage is set for a full-scale exploration of the ways in which this disarmingly simple one-dimensional instruction book is converted into the four dimensions of space and time that characterize living organisms." He puts his finger on the significance of the genomic revolution: Almost infinite complexity emerges from a molecular code of mind-boggling simplicity. It is not enough for the DNA to spin off proteins. It must spin off the right proteins at the right time and in the right place in the life cycle of an organism, and do so reliably throughout the life of the organism.

As I stand by the water meadow, all of that wonderful molecular machinery is spinning and weaving in every cell of my body, and in every cell of every creature in the singing green expanse before me. Even the thoughts in my head as I bend to attend to the grassy ooze depend upon an unceasing whirlwind of chemical activity that I cannot see or feel, and that until recently we knew nothing about. The genomic revolution has made it clearer than ever that "to the very root and foundation of his nature man is one with the rest of the organic world." And not just the organic world. The unity goes deeper, to the stony foundations of the planet itself.

MY DAUGHTER, the geologist, gave me three zebra stones for my birthday. As I opened the package, I thought I had been given beautifully decorated ceramic tiles. Certainly, the thin flat stones have the look of human artifacts: a zebra-stripe pat-

tern in colors of chocolate and sand, rich in variation yet pleasingly rhythmic. Each of my stones is
different, but I did not doubt that they came from
the same artistic hand. My daughter set me straight.
The "tiles" are natural. They were sawed from stone
quarried in the Kununurra district of western Australia, then polished to a high gloss. They consist of
fine-grained, 600-million-year-old siltstone or clay-
stone that has not been found in any other part of
the world. Geologists do not agree on the origin of
the curious stripes. The chocolate bands are colored
by iron oxide, and the lighter background is the
color of typical silt. Perhaps layers of differently colored mud were deposited in a zebra pattern as we
find them, under the influence of wind or water.
Perhaps fluid sediments slumped and interleaved
prior to solidification. Or perhaps water seeping
through the deposits selectively leeched out the
iron oxides. We may never know how nature made
the zebra stones, which is just as well, since it is
their apparently gratuitous beauty that makes them
so interesting.

When the silts were deposited, Australia was lo-
cated somewhere north of the Earth's equator, part
of a supercontinent known as Gondwana. The land
was bare of any green, and although it is not incon-
ceivable that microorganisms played some role in
coloring or leeching the deposits, it seems far more
likely that nature used for its art no other instruments than law and chaos. Law is built into the uni-

verse from the first moment of its creation, as propensities of matter and energy to behave in certain orderly ways. The laws of nature, which physicists seek, are presumably constant throughout the universe. Why such laws should exist, and why they should be discoverable by the human mind, are great mysteries. But nature is not *entirely* lawful; if it were, the universe would be a boring place. All events may be lawful at some fundamental level, but in practice, webs of causality are so multitudinous and tangled that it is impossible to predict the detailed outcome of any event. Chaos is an inevitable consequence of complexity. It might also be true, as many quantum physicists believe, that unpredictability is built into the very fabric of creation, tempering lawfulness with uncertainty.

Everywhere we look in nature, we see both lawful pattern and chaotic variation: in the iridescence of the dragonflies that skim the surface of the water meadow, in the lapidary nuance of beetles enthroned at tips of grass, in the song of the spring peepers. And in the stripes of my zebra stones. The geologist may one day find a satisfactory explanation for the zebra pattern; we might even come to understand why the human mind finds the stones beautiful. All we can presently say is that we perceive beauty in a balance of order and disorder—the themes and variations of a Bach fugue, for example, or the off-kilter symmetry of Michelangelo's *David*—perhaps because the human mind is itself a product of law and chaos. It is intriguing that the sinuous interfolded pattern of my zebra stones bears a superficial resemblance to the ancient Chinese symbol of yin and yang. Law and chaos are the dual creative principles of nature. No universe is possible without a generative friction between life and death, noise and silence, rigidity and fluidity, repetition and disruption, fire and ice. Here in the water meadow these strident polarities are all around me.

In one of his most popular essays, "The Colloid and the Crystal," the nature writer Joseph Wood Krutch wrote about these opposing forces in nature. "Order and obedience are the primary characteristics of that which is not alive," he wrote. "Life is rebellious and anarchical." He was wrong to identify obedience and rebellion with nonlife and life, respectively. We now know that the inanimate snowflake crystal, so apparently lawful and static, grows its six-pointed form under the controlling influence of exquisite molecular vibrations—every water molecule in the crystal is permanently aquiver, each in resonance with all the others—and it is delicate instabilities in these vibrations that allow snowflakes their lovely variations on the six-pointed theme. And life, we now understand, would not be possible unless nature had contrived elaborate molecular machinery to detect and repair any rebellious deviation of an organism's genetic code. The inanimate and the animate are equally products of law and chaos. Still, Krutch was right when he said that "the ultimate All is not one thing but two."

IGNORANCE IS NOT BLISS. But ignorance of ignorance can be bliss. Those who reside in the certainty of simple pieties are often happier than those like Peattie, Eiseley, and Thomas who toss and turn in their incertitudes. For the person who is willing to live with complexity, ignorance is a provocation, a goad, a confining screen to be pricked through. What we see when we prick through the screen is as often bedevilment as delight. Beyond the screen is another screen, and beyond that, another. See that static patch of white on the distant hillside, says the Roman philosopher Lucretius, in his poem *On the Nature of Things*, like a field of snow shining in the sun? Move closer. It is a flock of sheep, gamboling. Likewise, a grain of sand, examined more closely, might reveal its constituent atoms, guessed Lucretius, and of course he was right. The tiny six-pointed snowflake is, on a deeper level, a buzzing hive of molecular vibrations. And so, too, the lush diversity of life in the water meadow, examined more closely, resolves itself into a fandango of dancing molecules. The seen is a mask for the unseen. Our eyes open at birth to a flood of photons, but we must learn to see.

Lucretius could only intuit the existence of atoms. We have invented ingenious instruments that extend our vision into the realm of the atoms and the galaxies. Further, we know, as Lucretius did not, that the radiations to which the human eye is sensitive are but a narrow window on a spectrum of ambient energies ranging from long-wavelength radio waves to the short-wavelength gamma rays. All of these energies carry messages from the world. Slowly, painstakingly, we have devised the techniques that let us see things hidden to unaided human sight. And the more we see, the more we come to understand how much more remains unseen. This is the greatest and most challenging discovery: our abysmal ignorance.

In a poem titled "Seeing Things," the poet Howard Nemerov describes an experience he had standing by a water meadow.

When a silhouetted tree against the sun
Seemed at my sudden glance to be afire:
A black and boiling smoke made all its shape.

Such moments of epiphany are rare, no less for poets than for ourselves. In the climactic image of *Pilgrim at Tinker Creek,* Annie Dillard also encounters the "tree with the lights in it." She writes, "It was less like seeing than like being for the first time seen, knocked breathless by a powerful glance." The poet Sylvia Plath describes a similar experience in "Black Rook in Rainy Weather," a bird perched in a tree that can "seize my senses, haul my eyelids up, and grant a brief respite from fear of total neutrality." In each case, what has happened to the poet is not so much a perception as an intuition, an encounter with the quick of creation, a sudden immersion into mystery, a revelation of something behind the mask. We live, of course, in a sea of mystery, the great roiling ocean of our ignorance; we are up to our eyeballs in it, awash to the tops of our heads. But usually we are blinded by the mundane, straitjacketed by the familiar. Our senses are dulled by the tedium of the commonplace. Hungry for grace, we turn to the poets for intuitions of the terrible and sublime, for confirmation that the ordinary is not ordinary at all, that the commonplace is miraculous, that behind the mask of the water meadow is a deeper reality that consoles us in our yearning and desperation.

And so it was for Nemerov standing in his water meadow, observing the tree afire. But wait! He lifts his binoculars to his eyes, and the boiling smoke is resolved into a cloud of gnats, flitting in their millions. The tree's smoky aura is, after all, only bugs mating, the ceaseless machinery of life churning relentlessly on. It was, he says, as "Close as I ever came to seeing things / The way the physicists say things really are." Strike through the mask, you find another mask, he says. Like the patch of "snow" on the distant hillside that is actually a flock of sheep, or the water meadow that is a buzzing blur of spinning DNA, the apparently miraculous halo of the tree, closely observed, reveals itself to be a frenzy of insects:

copulatory organs, proboscises, antennae, wings. The black and boiling smoke is bugs.

THE AMES FAMILY once commissioned an attractive drawing of their family tree that was subsequently published on the cover of the *Boston Globe's* Sunday magazine section to illustrate a story about the family and our town. There's old Oliver standing firm as the trunk, bifurcating into the two thick branches of his sons, Oakes and Oliver II. The lineage branches again and again—grandchildren, great-grandchildren. Here in the fifth generation of leafy twigs in a full canopy of leaves are the Sheep Pasture children, Elise, Olivia, Oliver, and Richard. Imagine, if you can, a family tree that traced the ancestry of these children back through the millions of generations that separate them from the spring peepers and Canada geese of the water meadow. Imagine the gorgeously branching tree that includes ourselves, the red-winged blackbirds, and the crawfish in the slime. Go farther, back across the billions of generations to the microbial ancestors that would enable us to encompass within a single tree the respective lineages of the geese and the grasses, the entire familial relationships of the water meadow. Possible—barely—to imagine but not easy to execute.

Like my other grandchildren before her, my eighteen-month-old granddaughter Kate loves to visit the water meadow to feed the ducks and geese.

Like all kids, she is a fine taxonomist (one who classifies organisms into categories that reflect natural relationships). Canada geese and mallards are "quack-quacks." The dragonfly and mayfly are "bugs." Even a toddler who can barely talk recognizes that creatures have family relationships. She would probably disagree with professional taxonomists on some categories. Kate would surely lump eels with snakes, rather than fishes. A bat looks more like a bird than a puppy, although bats and puppies are biologically more closely related. Chimps and orangutans are "monkeys" to Kate, but it would never occur to her to put Mommy and Daddy—or even Grandpa—in the same category as chimps (primates, Hominoidea). For the professional taxonomist, neither chimps nor orangutans are true monkeys; scientifically speaking, humans, chimps, and orangutans are all apes. We all agree with Kate that creatures can be grouped into natural families, but not even professional biologists always agree what the families are. The inventory of classified living species currently stands somewhere near 2 million. The number of species awaiting classification may be 50 or 100 times greater. Finding the family relationships is a problem of surpassing complexity.

Constructing a family tree for the entirety of life takes a brave taxonomist, indeed. The British science journalist Colin Tudge tried it recently in his book *The Variety of Life: A Survey and a Celebration of all the*

Creatures That Have Ever Lived. He gives us the great pageant of life, microscopic and macroscopic, living and extinct, root and foundation, unfolding across the pages of his book from the hypothetical bacterial ancestors that mothered us all. The vast majority of species that have ever lived are extinct. For every Canada goose or mallard in the water meadow, there are about a hundred species of feathered creatures—*Iberomsornis, Enantiorithes, Patagopteryx, Archaeopteryx,* etc.—that we know only by their fossils. The pre-Darwinian view of life as a steady progress from primitive to complex, a great chain of being rising ultimately to—you guessed it—us, gives way in Tudge's account to a magnificent and richly proliferating tree on which the human species is but a single twig. Of course humans are a very untypical twig on the tree of life. The dazzlingly complex brains that burn atop our spines have changed forever the dynamic of life on Earth. No other large animal remotely approaches our numbers—6 billion strong and increasing, usurping more and more of the planet's resources for ourselves. There is no way human population can continue to grow without ensuring the extinction of many of the marvelous living creatures cataloged in Tudge's sweeping compendium.

How will we make the awesome decisions about which creatures we will take with us into the future? The AIDS virus? The malaria pathogen? Probably not. But what about the mosquito? The housefly? The elephant? The hummingbird? The Book of Job said it long ago: "Ask now the beasts, and they shall teach thee; and the fowls of the air, and they shall tell thee." Only an immersion in nature will give us the requisite wisdom. I stand with my granddaughter—a darling twig on the tree of life—by the side of the water meadow, tossing crumbs to the bold *Branta canadensis* (Canada geese) and the less aggressive *Anas platyrhynchos* (mallards) and feel the tug of billions of generations rooting me in mystery, in the foundational muck of the meadow, in the teeming dark abyss of time.

GARDENS

THE PATH RISES from the water meadow to an old orchard—two trees still standing—where tree swallows colonize bluebird boxes. The boxes aren't meant for swallows, but what beautiful birds they are! Shimmering blue-green backs. White bellies. They stitch the morning meadow, quicker than the eye can follow. I applaud their "airshow" acrobatics, admire their "daredevil" swoops—until I see them harassing bluebirds. Then scolding metaphors come to mind. The swallows are "feisty," I think. They dart with "furtive" speed. They fling themselves into the air like "sky-smart, teenage hoodlums looking for trouble." My sympathies clearly lie with the "more civil" bluebirds.

What is this compulsion to anthropomorphize? Why can't I just let the swallows and the bluebirds be? It's an old habit, I suppose, ingrained in our race, to see ourselves in the animals. Aesop did it. Philosophers and theologians of the Middle Ages did it. Disney did it. In a book that educated and entertained my grandparents, the early-twentieth-century American nature writer Neltje Blanchan described the house wren: "If you fancy that Jenny Wren, who is patiently sitting on the little pinkish chocolate-spotted eggs in the center of her feather bed, is a demure, angelic creature, you have never seen her attack the sparrow, nearly twice her size, that dares put his impudent head inside her door. Oh, how she flies at him! How she chatters and scolds! What a plucky little shrew she is, after all!" In another bird handbook from early in the last century, Mabel Osgood Wright described the American crow as a "feathered Uriah Heep" and the jay as a "Robber Baron." The bluebird is the "color-bearer of the spring brigade," and the song sparrow is "the bugler." My favorite bit of bird anthropomorphism is F. Schuyler Mathews's description of the meadowlark's song as the first two bars of Alfredo's aria in *La Traviata*, "sung with charming accuracy." Mathews was another of the ornithologists who instructed my grandparents. Styles have changed. Anthropomorphic references in a nature handbook today would undermine the author's credibility. We no longer see ourselves in the animals. The new fashion is to see the animals in ourselves.

Sociobiologists and evolutionary psychologists lead the way. They look to our affinity with other animals to explain not only our physical bodies but also our behaviors. E. O. Wilson, the father of sociobiology, writes that "the brain exists because it promotes the survival and multiplication of the genes that direct its assembly. The human mind is a device for survival and reproduction, and reason is just one of its various techniques." So we look into our brain stems for shadows of reptilian ancestors. We examine the anthill for the origins of human societies. We watch gorillas in the mist to discover the roots of human aggression and sexuality. Even our ethical systems and religions might be part of our evolutionary heritage, say sociobiologists; they helped us survive and reproduce in chancy environments.

It is not so much that the crow is a "feathered Uriah Heep," they say, as that we have something of the crows within ourselves. It is not so much that jays are "Robber Barons" as that human robber barons share self-aggrandizing tendencies with jays. This inversion of viewpoint changes the way we think about environmentalism, ecology, and animal experimentation, perhaps even how we eat. Philosophers and theologians have yet to work out how it effects our sense of human purpose, but this much is clear: We are not angels burdened with animal bodies; we are animals who aspire to be angelic.

BEYOND THE FIELD where the tree swallows play is an acre or two of land set aside by the Natural

Resources Trust of Easton for community gardening. Any citizen of the town can claim a plot. Each
spring the individual plots are rototilled by the Trust.
Then one by one the gardeners appear with their
packets of seed. All summer long they weed and till
and shoo away birds. In autumn they harvest. They
are sometimes there in the early morning as I pass
on my way to work, wielding a hoe or down on their
hands and knees, tending their bounty. Onions.
Radishes. Lettuce. Tomatoes. Beans. Corn. There is
no real need for any of this. The gardeners could
buy their vegetables at the store for less money than
they spend on garden tools. But money is not the
point. What's really going on is a love affair with
seeds, with soil, with the sweet tactile sensation of
snapping a homegrown sugar pea or holding a hefty
tomato in the hand. The gardeners are there for
pretty much the same reason I walk the path rather
than drive to work, the same reason Bob Benson is
over there by the bluebird boxes checking nests. The
gardens are a way of cultivating a sense of place.

"Stay away from anything that obscures the place
it is in," writes poet Wendell Berry. He is not, I
think, asking us to leave landscapes alone; we no
longer have that privilege. When our first human
ancestor crafted a chopper out of stone and struck a
fire with flint, untrammeled nature was in retreat.
Frederick Law Olmsted employed thousands of
men and moved countless tons of earth to make

that patch of "natural" place called Central Park; he reshaped the landscape, to be sure, but in a way that lets place shine through. Even such ostensibly wild places as Yosemite and Acadia National Parks show the marks of his civilizing influence. Part of the requirement for the design competition for New York's Central Park was provision for cross-town traffic; after all, the park was to extend fifty-one blocks up the center of Manhattan Island and could not be a vehicular barrier for such a distance. Olmsted solved the problem by sinking transverse roads in deep-walled trenches, thereby preventing carriages (and later automobiles) from obscuring the visual integrity of the park, a strategy that minimizes the influence of vehicular traffic even today. Wendell Berry, that poet champion of cherished places, is also a farmer who knows that a dairy cow and an ear of corn are artifacts, bred by human artifice from wild species. A farm or a garden is an artifact. The question is not whether we will live in artificial places but—and this is surely his point—whether the places we live in encourage a sense of belonging.

If we don't belong somewhere, we belong nowhere. If we are not attached to a particular landscape, we might as well be adrift in space. The Alaskan nature writer Richard Nelson says: "What makes a place special is the way it buries itself inside the heart, not whether it's flat or rugged, rich or austere, wet or arid, gentle or harsh, warm or cold, wild or tame. Every place, like every person, is elevated by the love and respect shown toward it, and by the way in which its bounty is received." The place we learn to love can be a windowsill in a New York high-rise, a patch of woods on Walden Pond, or a million acres of the high Sierras. What's important is that we feel at home.

Perhaps one reason Frederick Law Olmsted was so successful at creating landscapes that make us feel at home is that he designed them at his own gracious home, Fairsted, in Brookline, Massachusetts. The house and grounds are today in the care of the National Park Service, a pocket paradise within a short trolley ride of downtown Boston. The house, by no means grand, has rooms for the offices where Olmsted's designers plied their trade, startlingly small; it is astonishing to think how dramatically what happened in these few cramped spaces transformed the landscape of America. (The park systems of Atlanta, Baltimore, Boston, Buffalo, Chicago, Denver, Louisville, and Seattle are just a few of the Olmsted firm's many creations.) The grounds of Fairsted were a kind of minilaboratory for Olmstedian ideas. Here were trained some of the most prominent landscape architects of the late nineteenth and early twentieth centuries: John Charles Olmsted, Frederick Law Olmsted Jr., Charles Eliot, Henry Sargent Codman, Warren Manning, Arthur Shurcliff, James Frederick Dawson, and Henry Vincent Hubbard. Perhaps never in the history of the human race has such a civilizing influence emerged from so compact an engine of change.

Imagine what our cities and suburbs might be today if those in charge of the planning and execution of public and private development were guided by Olmstedian principles. Instead, we create landscapes that cater to automobiles, not people, even to the point of sacrificing the integrity of some of our forbears' most precious gifts, such as Charles Eliot's great system of metropolitan parks and parkways around Boston or Connecticut's graceful Merritt Parkway. As early as the 1920s the writing was on the wall. On September 29, 1923, Charles Eliot's friend and coworker Sylvester Baxter wrote in the *Boston Evening Transcript*, "The parkways and boulevards . . . intended to be strictly subordinate . . . to make the reservations pleasantly and easily accessible . . . have become the primary factor in the scheme of the park system." The service of motor traffic had become the main consideration of the park administration, he complained. The automobile is here to stay, of course, and rightly so, but we are not required to love it, or sacrifice everything to it. Every acre of unnecessary asphalt is one less natural place to love. "There are no unsacred places," says Wendell Berry, "there are only sacred places and desecrated places."

I count myself fortunate that my path takes me through a landscape that has been preserved from desecration. In the community gardens I encounter others who share that good fortune, and who cultivate a sense of place along with their corn, beans, and radishes. The gardeners hail me as I pass. Sometimes they offer me produce, a few radishes perhaps; sometimes I buy bouquets of flowers from jam jars at a pathside stand maintained by the Natural Resources Trust; one dollar in the box, honor system. Bob Benson, the bluebird man, never fails to share with me his knowledge as I pass. Many mornings I've soaked my shoes wading into the dewy grass beyond the path to inspect a new bluebird box, perhaps of Bob's own design. At the Trust's annual Harvest Fair, held on the site of the Sheep Pasture mansion in early October, our local wild bird emporium solicits sponsorships for bluebird boxes. Bob makes sure the boxes will appeal to breeding bluebirds.

WITH BOB I have watched bluebirds pass in a few weeks from egg to fledgling, or with the gardeners watched a tomato seed become in the course of a summer a tall plant burdened with fruit. We are witnessing invisible genetic scripts assemble atoms gleaned from soil and air into birds or vegetables. The earliest cells in the process, the so-called stem cells, have the potential to become any other kind of cell—feather or beak, stem or leaf. What they become depends upon which genes are expressed at each stage of development, and that in turn is sensitive to the environment. The DNA is not quite a blueprint, because there's not a building contractor to follow the plan. Neither is it quite a computer program, because there's no hardware to run it. Here's how geneticist Enrico Coen puts it in his book, *The Art of Genes*: "The software, the program, is responsible for organizing hardware, the organism. Yet throughout the process, it is the organism in its various stages of development that has to run the program. In other words, the hardware runs the software, whilst at the same time the software is generating the hardware." If that sounds circular, it is because it is. The essence of life is circularity. It's a bootstrap process. The bluebird and the tomato pull themselves into being. So do you and I.

My 165-pound body consists of about 110 pounds of oxygen, 30 pounds of carbon, 16 pounds of hydrogen, 6 pounds of nitrogen, and 3 pounds of everything else. Basic stuff, mostly, the stuff of water and air. You'd think we could get almost everything we need to build our bodies by taking deep breaths and gulps of water. But it's not quite that simple. Consider those 6 pounds of nitrogen in my body. Our cells build proteins by stringing together chemical units called amino acids, and every amino acid contains a nitrogen atom. Without nitrogen, no proteins. Without proteins, no me—and no tomato or bluebird; our bodies are tissues of proteins. So what's the problem? The atmosphere is

80 percent nitrogen. We suck in a lungful of nitrogen with every breath. But the nitrogen in the atmosphere (and in our lungs) is useless. The two nitrogen atoms in a nitrogen molecule are so tightly bound together that they are essentially inert; they hardly react with anything else. This inertness of nitrogen would seem to be a flaw in nature's Tinkertoy set, at least as far as life is concerned. We live in a sea of nitrogen, and it does us no good at all.

At least not directly. Since we cannot use nitrogen from the air, our bodies extract nitrogen atoms from less tightly bound molecules in the food we eat, and from these we make amino acids and ultimately proteins. Even then, there are ten amino acids that we can't manufacture ourselves—the so-called essential amino acids—and for these we must rely upon plants, which alone have the ability to make all twenty kinds of amino acids that our bodies require. Without plants—without those ten essential amino acids that animals cannot make—you and I could not exist. And where do the plants get *their* nitrogen to make amino acids? Bacteria that live with certain plants have the ability to do what we can't do and what the plants themselves can't do: take nitrogen from the atmosphere, break those devilish bonds, and incorporate the nitrogen into molecules that plants and animals can use. This happy alliance is called *nitrogen fixation*. The bacteria get energy from the photosynthesizing plants; the plants get useful nitrogen for building amino acids, including the ones we require. So ultimately, the whole pageant of life on Earth depends upon bacteria that live in or around the roots of certain plants.

But wait! What I just said about our dependence upon nitrogen-fixing bacteria is not strictly true. It might have been true a century ago, but it is not true today. Humans have a way of taking matters into their own hands. In 1909 a German chemist named Fritz Haber invented a way to use high temperatures and pressures in the presence of a catalyst to make atmospheric nitrogen react with hydrogen to form ammonia: artificial nitrogen fixation. Like bacteria, the Haber process breaks the bonds of atmospheric nitrogen and renders it useful as fertilizer for agriculture; it also breaks the bonds that have made us dependent upon nitrogen-fixing bacteria since the dawn of time. Of course, as usual when we mess with long-established patterns of life, we create problems, too. The Haber process uses huge amounts of unrenewable fossil fuels as its source of energy, and the runoff of excess artificial fertilizers from fields and gardens poisons lakes and streams. What at first seemed a blessing of technology has come at considerable environmental cost. But the melancholy fact is that because of overpopulation the human need for food now far outstrips the ability of bacteria to supply us with nitrogen. More nitrogen fertilizer is today applied in agriculture than is fixed naturally by bacteria in all terrestrial ecosystems. Almost all nitrogen in the fields of

Egypt, Indonesia, and China comes from synthetic fertilizer—100 million tons of it a year. If it weren't for the Haber process, lots of people would be starving. Or to put it another way, if it weren't for the Haber process, there wouldn't be so many of us. Pollution or starvation; a stark choice.

In the community gardens the problems of feeding the global billions seem far away. Here the goal is to let bacteria do the work of nitrogen fixation and avoid artificial fertilizers. The gardeners plunge their hands into the soil and celebrate an organic intimacy with the ancient miracle of sun, seed, leaf, root—and those unseen nitrogen-fixing microbes that make it all possible. But of course the notion of "organic" gardening is illusory in several ways. A garden is by definition artificial. The plants are the result of thousands of years of selective breeding. The gardeners arrive at their "organic" plots in machines driven by fossil fuels. Their gasoline-powered rototillers grind away, displacing birdsong in the morning air. The entire landscape of the path is a work of human design, by and large splendidly successful, but artificial nonetheless.

TEN THOUSAND YEARS AGO, humans learned how to farm. It was an epochal invention that made possible settled life, cities, craft specialization, writing, organized religion, architecture, mathematics, science. Now humanity stands on the brink of a second agricultural revolution potentially as great as the one that occurred when our ancestors gave up a hunter-gatherer way of life and settled down as farmers. Scientists and engineers are poised to genetically modify organisms to increase the yield, nutrition, freshness, and pest resistance of food plants and animals, and perhaps even to diminish the use of artificial fertilizers (and fossil fuels) by supplementing biotic nitrogen-fixation systems. Other possible benefits of genetically modified (GM) organisms include improved use of marginalized land—saving wild areas from the plow—and abundant production of vaccines and pharmaceuticals, possibly eliminating diseases such as cholera, hepatitis B, and malaria. The promise is great. But as always with the products of human artifice, not without attendant dangers.

Humans have been genetically modifying plants and animals for millennia, by selective breeding. For example, as biologist Richard Lewontin has pointed out, the large kernels of domesticated corn are ideal for harvesting and storing, which is why the plant was developed, but such a plant would soon disappear from nature without human cultivation because it could not disperse its seeds (which are too big and heavy to be carried by wind). In other words, the corn that is so carefully cultivated by "organic" gardeners is itself an unnatural contrivance of human ingenuity. But there *is* a difference between selective breeding and genetic engineering. Till now, all plant and animal modification has been achieved by crossbreeding between closely related species. The new twist is that scientists have the power to *mix* the genes of organisms, no matter how dissimilar. Bacterial genes have been inserted into corn, for example, conferring pest resistance. Firefly genes have been inserted into tobacco plants, causing them to glow (mainly to test techniques of genetic transfer). Even monkeys have been made to glow with the genes of luminous jellyfish.

No technology has excited more awe and apprehension than this ability to manipulate the code of life and jumble species. Olmsted moved millions of tons of earth to make landscapes more "natural" than nature had achieved on its own; the genetic engineers move bits of DNA to create organisms more perfectly suited to human use than did 4 billion years of haphazard evolution. As I write, more than 100 million acres in the United States are planted with genetically modified plants: soybeans, corn, squash, potatoes, canola, and sugar beets. Most commonly, these plants are modified with bacterial genes that make them more resistant to insect pests or to weed-control herbicides. It is not yet exactly clear who benefits by this development— agribusiness, farmer, consumer—but the trend is certain to continue. So far, there is no case of GM crops proving harmful to humans. At the same time, the long-term safety of GM foods—for consumers and the environment—has not been established. My own suspicion is that a century from now GM foods will be taken for granted, as one component of a yet-to-be-imagined ecological agriculture, incorporating new crossbred crops, mixed-crop farming, crop rotation, and the sparing use of synthetic fertilizers and pest controls.

It is hard to think of any new technology that has not had the potential for harm. Thoreau protested the Fitchburg Railroad that ran hard by Walden Pond; in the time it took him to earn the fare, he said, he could walk to Fitchburg with more pleasure. Ames shovels helped build the railroads that bound America together into one strong nation; the company also made entrenching tools for the Union soldiers who fought and died by the tens of thousands in a cruel and bitter civil war. Few people would choose to reverse the technological clock, even for a generation, much less for the thousands of years that separate us from our pretechnological ancestors. Thoreau's brother John died of sepsis from a shaving nick; modern drugs would have saved his life. The thud of stream-driven hammers resounded in the ears of nineteenth-century North Easton villagers ten to twelve hours a day; that awful din created the wealth that enables their twenty-first-century successors to potter in community gardens and tend bluebird boxes. Ames wealth preserved the green space of the path that gives present-day villagers access to nature.

A busy highway separates the community gar-

dens of Sheep Pasture from Stonehouse Hill
House on the carefully sculpted former Ames estate that is now the campus of Stonehill College.
My one-mile trek has taken me to the beginning
of the universe and back, to the core of the Sun
and the spinning heart of DNA; it has also led me
to the threshhold of a new millennium. What sort
of global environment awaits my grandchildren
and great-grandchildren? Will planetary evolution
recapitulate the history of the path: wilderness,
hunter-gathering, agriculture, industry, cybercul-
ture? Will third-world poverty and environmental
despoliation give way in time to the local equivalent of bluebird boxes and community gardens?
We are a species of animal that evolved in a particular environment—the savannas of East Africa—
with behaviors and predispositions shaped for
optimal survival in that environment, and here
along the path has been recreated something resembling that place, but with the comforts and se-
curities of technological civilization. Hollywood
and sci-fi writers often evoke a grim inorganic future, all gleaming steel or rusted iron, inhabited
by techno-mutants; doom-and-gloom environmentalists envision a future wasteland of asphalt,
pollution, and creaturely solitude. But what I've
seen along the path suggests otherwise. Given a
modest level of affluence and the freedom to
choose, most of us will opt for a garden planet—

an Arcadia somewhere between the poles of the wild and the artificial.

THE PATH WINDS beneath an overarching canopy of trees, then opens to the dawn. It is one of those brisk fall mornings when the sky turns a Maxfield Parrish blue just before sunrise. One, two, three ragged files of Canada geese skim the treetops, preceded and followed by their honking chorus. I freeze in my tracks to watch them pass, heading south, feathers ruffled by the last warm breezes of the season. When their honks have faded into silence, I notice a chill in the air. The spinning planet has leaned into its winter curve, away from the Sun. And then, just when I think the racket has passed, I hear another barely audible chorus of honks, high in the air. I look up to see a long, asymmetrical vee of perhaps a hundred geese, moving south at high altitude, catching the direct rays of a Sun that has not yet broken the horizon, like gold doubloons spilled across the sky.

Why the vee? Why that Euclidean arrow aimed at warmer climes? It is a fundamental tenet of science that things don't happen by happenstance, or merely to strew a vee of gold across a morning sky for the benefit of a human watcher. We find vee formations beautiful, but our taste for geometry has no value as explanation. Adaptation is the key to understanding the vee. Formation flying may increase aerodynamic efficiency by giving each successive goose in line a bit of extra uplift with something aeronautical engineers call *wingtip vortex*. Or geese may fly in vees merely to keep each other in view. Whatever the reason, creatures inevitably evolve the competitive edge that helps their genes flow into the future. It was Darwin's genius to recognize that natural selection leads to complexity—even, in the case of the Canada geese, to a spectacular morning geometry. The law of entropy asserts that as time passes the universe moves—on balance—toward disarray. Life builds patterns of order by drawing upon a correspondingly greater *disordering* process at the center of the Sun where solar energy is produced. The geese that etch their vees against the morning sky are creatures of our yellow star, which now—just now—lifts its fiery rim above the horizon, bathing the path in golden light.

We, too, enhance our genetic capital at the expense of our star. We are, all of us, building pinnacles of order in a universe that is destined, ultimately, to tumble all our towers to dust. But this is not reason for despair. The climb of life toward complexity has been going on for billions of years and will go on for billions more, as long as the Sun continues to shine. Because of its monumental selective advantage, the human brain, or something like it, was probably inevitable in the course of evolution, and with intelligence and self-awareness came science and technology. Knowledge once gained cannot be unlearned, and knowledge is power. For better or worse, the future of the planet has been handed to us, not by a deity but by fate. Stewardship of other creatures is in our hands. There is no transcendent moral imperative that directs us to contrive one future rather than another, or to preserve one creature rather than another. As awful as it sounds, our species' self-interest may be the soundest basis for choice. Just as the Golden Rule—do unto others as you would have them do unto you—acknowledges that individual self-interest is bolstered by altruism toward our fellow humans, so does an understanding of the ecological wholeness of the Earth suggest that our altruism should extend to other creatures, too: plants, animals, even microbes.

Environmental conservation—clean water and air, a steady climate—is in the interest of our species. Like Olmsted's parks, the Arcadia we should cultivate would have its "The Wilderness," "The Green," "The Deer Paddock," "The Long Meadow," "The Nethermead," and "Lookout Hill." It would be urban, but eschewing sprawl. It would have a place for automobiles, but would not surrender to them. It would admit genetically engineered plants and animals where these favor full bellies, health, clean water, and wild places. It would be utopian in vision, but realistic in its recognition that self-interest can be too narrowly conceived. It would understand that "nature" is more than the happy haunts of the Bobb-sey Twins; nature also includes bacteria, galaxies, quarks, and, yes, human culture. It would recognize that nature "in the raw" is built upon a platform of death and sometimes (by human standards) un-speakable cruelty, but that human moral choice can transcend the tooth-and-claw imperative of evolu-tion. The landscape architects of Olmsted's era thought they were improving nature with shovel, ax, and barrow, and perhaps they were. They sought aesthetic and spiritual values in their landscapes, and these are what I have sought along the path.

The Arcadian ideal of humans living in harmony with tamed nature did not begin with Frederick Law Olmsted, Capability Brown, or even the supposed Peloponnesian paradise itself (witness the more an-cient myth of the Garden of Eden), nor was it dis-credited by the obscenities of the twentieth century's wars, the Great Depression, or the grimmer excesses of technology. It is a sturdy old myth, and in it we might still hope to combine the Enlightenment, with its confidence in the power of the human mind to make sense of the world, and romanticism, with its belief that all of life is a miracle. Along the one-mile walk of the path, I have found these ostensibly com-peting tendencies happily fused: order and surprise, artificial and natural, civilized and wild, human self-interest and organic wholeness.

EPILOGUE

IN A PARK in West Bridgewater, about ten miles from North Easton, stands an old iron anvil. A plaque on a nearby forge-stone reads:

> The land of this park was bought in 1649 from the Massasoit Indians by Miles Standish and others as part of the Bridgewater Purchase and allotted to John Ames, an original shareholder and settler. And here before the Revolutionary War the fourth inheritor, Captain John Ames began the manufacture of shovels with a trip-hammer set on this stone.

These were the first iron shovels made in the colonies. It was Captain John Ames's son, Oliver, who moved the shovel works to North Easton in 1803 and built the company into the largest shovel manufactory in the world. Charles Carroll of Carrollton, the last surviving signer of the Declaration of Independence, used an Ames shovel to turn the first spadeful of earth for America's first railroad.

When the Ames Shovel Company closed up shop in North Easton and moved to West Virginia, in the mid–twentieth century, the handsome stone-and-timber buildings were purchased by developer Arnold Tofias and converted to offices, warehouses, and light industrial use. On the upper floor of the old Ames administration building were stored 1,500 linear feet of company documents: production records, sales ledgers, employee time books, payrolls, legal papers, inventories, and correspondence. In addition to this paper trove, there were almost 800 shovels, some dating back to the earliest days of the company. All of this material was donated to Stonehill College by Tofias, where it now resides as the Arnold B. Tofias Industrial Archives, one of the nation's best collections of primary source materials for industrial history. The shovels themselves are a vivid record of nation building: shovels for earth or coal, trenching spades, nursery spades, potato scoops, coffee bean scoops, snow shovels, shovels for draining fields, military trenching tools, garden tools, even toy shovels for children. The most spectacular part of the collection is a magnificent floor-to-ceiling, wood-and-glass case containing several dozen silver-plated Ames shovels that was exhibited at the 1876 Centennial Exposition in

Philadelphia, celebrating 100 years of the nation's history—and this is as good a place as any for the path to end.

The shovels in the centennial collection combine beauty and utility. The wooden handles extend and amplify human force; the steel blades slice and cut and carry. What makes these tools so attractive is their scale, their satisfying heft in the hands, the way they move in harmony with human limbs, the polish of sweat on grip and shaft. With these instruments nineteenth-century Americans turned a raw continent into a free, unified, and stable democracy, the most powerful on Earth. This achievement did not come without environmental devastation, political scandal, social injustice, and genocide. But the balance sheet of history will show, I believe, that a larger percentage of us live better, healthier, and longer lives than ever before.

About half of the Earth's land surface is presently exploited by humans, and all of the land and water surface is touched in some way by the waste products of human cunning. Energy use has grown sixteenfold in the last century, much faster than population. An invisible worldwide web of information dances between the planet's surface and orbiting satellites at the speed of light. As I write these final words, I am far from the path, on a little island in the tropics, but with a few taps of a finger I can retrieve on my laptop computer screen a satellite image of the path from space, on which I can recognize the woods, fields, bridge, and gardens. The chemist Paul Crutzen has proposed the term *Anthropocene* (*anthro* = human) to represent the present age of Earth's history, in which human artifice is the dominant geologic force. The Anthropocene can be said to have started in the latter part of the eighteenth century, he says—analysis of air trapped in arctic ice 200 years ago shows the beginning of growing global concentrations of carbon dioxide and methane. This date coincides with James Watt's innovations in steam power, and with the beginning of shovel manufacturing by John Ames in West Bridgewater.

The technological products of human ingenuity represent an inevitable stage in planetary evolution, yet our Arcadian yearnings are dictated by millions of years of pretechnological human evolution. It is a conundrum of human life that our intellects have outraced our instincts; cultural evolution has overtaken organic evolution. Biologically, we are hunter-gatherers who suddenly find ourselves in command of almost unimaginable powers for planetary transformation. We struggle to bring together our genes and our aspirations, Wilderness and Technopolis, the romantic and the visionary, spirit and economics. Scientists and engineers are responsible for ensuring that the Anthropocene Era will be good for the human race and good for the planet, with its diversity of creatures and habitats. Architects and planners are implicated, too, and the managers and stockholders of multinational corporations, politicians, philosophers, poets, and religious leaders. Most of us, however, will make our contribution for good or ill on the local scale, along paths that begin at our own front door.

NOTES

Prologue

3   "with reverent feet": Robert Lloyd Praeger, *The Way That I Went* (Dublin: Allen Figgis, 1980), p. 2.

4   "adequate step" and "To forget the dimensions": Tim Robinson, *Stones of Aran* (Dublin: Wolfhound Press, 1986), pp. 12–13.

5   "If you know one landscape well": Anne Michaels, *Fugitive Pieces* (New York: Knopf, 1997), p. 82.

Village

17   "wild musical sounds" and "unwearied and unresting": Henry David Thoreau, *A Week on the Concord and Merrimack Rivers* (Princeton: Princeton University Press, 1980), p. 116.

Woods

27   "a green Thought in a green Shade": Andrew Marvell, *The Poems and Letters of Andrew Marvell* (Oxford: Clarendon Press, 1952), p. 48.

33   "as full as she could stow": Governor William Bradford, quoted in William Cronon, *Changes in the Land: Indians,*

Colonists, and the Ecology of New England (New York: Hill and Wang, 1983), p. 109.

38 "was little more than a glorified gardener": S. B. Sutton, *Charles Sprague Sargent and the Arnold Arboretum* (Cambridge: Harvard University Press, 1970), p. 28.

39 "While all his surroundings": John Muir, *Atlantic* (July 1903), quoted by Sutton, p. 151.

41 "It is becoming rarer every year": Neltje Blanchan, *Wild Flowers* (New York: Doubleday, Doran & Co., 1917), p. 34.

44 "a more refined kind of nature": Andrew Jackson Downing, *Horticulture* (July 1848), quoted in Peter J. Schmitt, *Back to Nature: The Arcadian Myth in Urban America* (New York: Oxford University Press, 1969), p. 56.

46 says the last bear was killed: William L. Chaffin, *History of Easton* (Cambridge: John Wilson & Son, 1886), p. 17.

Rocky Foundation

55 has a theory about the ubiquitous stone walls: Robert Thorson, *Stone by Stone: The Magnificent History of New England's Stone Walls* (New York: Walker & Company, 2002).

59 "plainly once the work of man": Henry Vincent Hubbard and Theodora Kimball, *Introduction to the Study of Landscape Design* (New York: MacMillan, 1938, 1st ed., 1917), p. 69; quoted in Schmitt, p. 69.

60 "the beautiful and reposeful sights and sounds": Charles Eliot, as quoted in Hubbard and Kimball, p. 1; quoted in Schmitt, p. 66.

60 "Life, Power, Beauty, Peace": Frank Waugh, *The Natural Style in Landscape Gardening* (Boston: R. G. Badger, 1917), pp. 55, 59; quoted in Schmitt, p. 60.

60 "relief from the too insistently man-made surroundings" Frederick Law Olmsted, "Landscape Design," in *Significance of the Fine Arts* (Boston: Marshall Jones, 1923), p. 329; quoted in Schmitt, p. 57.

64–65 "Here is a community": Charles William Eliot, *Charles Eliot, Landscape Architect* (Boston: Houghton Mifflin, 1902), pp. 35–352; quoted in Norman T. Newton, *Design on the Land: The Development of Landscape Architecture* (Cambridge: Harvard University Press, 1971), p. 324.

67   "rapidly vanishing for all eternity": Frederick Law Olmsted Jr., quoted in Victor Shelford, ed., *Naturalist's Guide to the Americas* (Baltimore: Wilkins and Wilkins, 1926), p. 8; quoted in Schmitt, p. 154.

Verges

72   She might have used as her guide: Mrs. William Starr Dana, *How to Know the Wild Flowers* (Boston: Houghton Mifflin, 1989).

73   "'Some of these days,' wrote Burroughs": Quoted in Dana, p. ix.

76   "sprung up since the English": Quoted in Cronon, p. 143.

77–81   In this saga of two-way traffic: The Jefferson-Buffon story is recounted in Nathan Schachner, *Thomas Jefferson: A Biography* (New York: Thomas Yoseloff, 1951), pp. 284–86.

81   Mrs. William Starr Dana rose to rebut: Mrs. William Starr Dana, *According to Season* (Boston: Houghton-Mifflin, 1960), p. 149.

86   "Most young people find botany a dull study": John Burroughs, quoted in Dana, *How to Know the Wildflowers*, p. xv.

86–88   The physicist Eric Chaisson tries to answer these questions: Eric Chaisson, *Cosmic Evolution: The Rise of Complexity in Nature* (Cambridge: HarvardUniversity Press, 2001).

91   "That we know so little": Dana, *According to Season*, p. 3.

Brook

94   "Love, we are a small pond": Maxine Kumin, *Up Country* (New York: Harper & Row, 1972), p. 61.

97   "Our oceans were once our rocks": P. W. Atkins, *Molecules* (New York: Scientific American Library, 1987), p. 23.

99   It is reproduced in: Hazel Varella, *Growing Up at Sheep Pasture* (Easton: Easton Historical Society, 1976).

100   As historian Peter Schmitt has described: Schmitt, p. 115.

104   Historians of science: Giorgio de Santillana and Hertha von Dechend, *Hamlet's Mill: An Essay on Myth and the Frame of Time* (Boston: Gambit, 1969).

105   "If the stars should appear one night in a thousand years": Ralph Waldo Emerson, *The Collected Works of Ralph Waldo Emerson*, Vol. I (Cambridge: Harvard University Press, 1971), p. 8.

Open Fields

111   "Ivy-berries ripe": This and other quotes are from Gilbert White, Walter Johnson, ed., *The Journals of Gilbert White* (Cambridge: M. I. T. Press, 1970), p. 161.

111   a little book that remains in print today: Gilbert White, *The Natural History of Selborne* (London: Oxford University Press, 1971).

114   One human character: White, *The Natural History of Selborne*, p. 200.

116   "the least insect or animal": Walt Whitman, *Complete Poetry and Collected Prose* (New York: Library of America, 1882), p. 171.

116   "The nearest gnat is an explanation": Walt Whitman, p. 243.

120–21   "It was a secret and ardent stirring": Thomas Mann, *The Magic Mountain* (New York: Alfred A. Knopf, 1927), p. 350.

124   "At bottom, the whole concern of": William James, *The Varieties of Religious Experience* (Cambridge: Harvard University Press, 1985), p. 41.

124   "for everything which is natural": E. E. Cummings, George J. Firmage, ed., *Complete Poems 1904–1962*, (New York: Liveright, 1991), p. 663.

126   "She was questioning not only": Paul Brooks, *The House of Life: Rachel Carson at Work* (Boston: Houghton Mifflin, 1972), p. 293.

129   "to a deep genetic memory": E. O. Wilson, *Biophilia* (Cambridge: Harvard University Press, 1984), p. 111.

131   "into harmony with the age's humanism": Quoted in Rebecca Solnit, *Wanderlust: A History of Walking* (New York: Penguin, 2000), p. 91.

131   "th' enchanted round I walk": Quoted in Solnit, p. 92.

NOTES

Water Meadow

133 "The idea [of the helix] was so simple": James Watson, *The Double Helix* (New York: Atheneum, 1968), p. 114.

138 "the trees blasted by the great guns": Donald Culross Peattie, quoted in *The Norton Book of Nature Writing*, Robert Finch and John Elder, eds. (New York: W. W. Norton & Co., 1990), p. 449.

138 "a man unusually well trained": W. H. Auden, from introduction to Loren Eiseley, *The Star Thrower* (New York: Times Books, 1978), p. 20.

139 "I say that it touches a man": Donald Culross Peattie, quoted in *The Norton Book of Nature Writing*, p. 456.

140 "It was the equivalent of finding": Adrian Desmond, *Huxley* (Reading, Mass.: Addison-Wesley, 1997), p. 218.

141 "To the very root and foundation": Quoted in Desmond, p. 241.

141–42 "The stage is set for": Quoted in *Nature*, 408: 894–96 (2000).

145 "Order and obedience": Joseph Wood Krutch, quoted in *The Norton Book of Nature Writing*, p 446.

147 "When a silhouetted tree": Howard Nemerov, *The Collected Poems of Howard Nemerov* (Chicago: University of Chicago Press, 1977), p. 479.

147 "It was less like seeing": Annie Dillard, *Pilgrim at Tinker Creek* (New York: Harper's Magazine Press, 1974), p. 33.

147 "seize my senses": Sylvia Plath, *The Collected Poems* (New York: HarperPerennial, 1992), p. 56.

150–51 The British science journalist Colin Tudge: Colin Tudge, *The Variety of Life: A Survey and a Celebration of All the Creatures That Have Ever Lived* (Oxford: Oxford University Press, 2000).

152 "Ask now the beasts, and they shall teach thee": Book of Job 12:7, King James Version.

Gardens

154 "If you fancy that Jenny Wren": Neltje Blanchan, *Birds* (New York: Doubleday, Doran & Co., 1917), p. 46.

154 "feathered Uriah Heep": Mabel Osgood Wright, *Birdcraft* (New York: Macmillan, 1936), p. 179.

154 "sung with charming accuracy": F. Schuyler Mathews, *Field Book of Wild Birds and Their Music* (New York: G. P. Putnam's Sons, 1904), p. 59.

155 "the brain exists": E. O. Wilson, *On Human Nature* (Cambridge: Harvard University Press, 1978), p. 2.

156 "Stay away from anything that obscures": Wendell Berry, *Poetry* (January 2001), p. 270.

157–58 "What makes a place special": Richard Nelson, *The Island Within* (New York: Vintage Books, 1991), p. xii.

159 "The parkways and boulevards": Quoted in Newton, p. 333.

159 "There are no unsacred places": Berry, p. 270.

161 "The software, the program, is responsible": Enrico Coen, *The Art of Genes* (Oxford: Oxford University Press, 1999), p. 11.

Epilogue

176 Paul Crutzen has proposed: Paul J. Crutzen, *Nature* 415: 23 (2002).

INDEX